# Conference Proceedings of the Society for Experimental Mechanics Series

*Series Editor*

Kristin B. Zimmerman, Ph.D.
Society for Experimental Mechanics, Inc.,
Bethel, CT, USA

More information about this series at http://www.springer.com/series/8922

Luciano Lamberti • Ming-Tzer Lin • Cosme Furlong • Cesar Sciammarella
Editors

# Advancement of Optical Methods in Experimental Mechanics, Volume 3

## Proceedings of the 2017 Annual Conference on Experimental and Applied Mechanics

Springer

*Editors*
Luciano Lamberti
Politechnico di Bari
Bari, Italy

Cosme Furlong
Mechanical Engineering Department
Worcester Polytechnic Institute
Worcester, MA, USA

Ming-Tzer Lin
National Chung Hsing University
Graduate Institute of Precision Engineering
Taiwan, China

Cesar Sciammarella
Illinois Institute of Technology
Chicago, IL, USA

ISSN 2191-5644 ISSN 2191-5652 (electronic)
Conference Proceedings of the Society for Experimental Mechanics Series
ISBN 978-3-319-63027-4 ISBN 978-3-319-63028-1 (eBook)
DOI 10.1007/978-3-319-63028-1

Library of Congress Control Number: 2015955141



Printed on acid-free paper

This Springer imprint is published by Springer Nature
The registered company is Springer International Publishing AG
The registered company address is: Gewerbestrasse 11, 6330 Cham, Switzerland

# Preface

*Advancement of Optical Methods in Experimental Mechanics* represents one of nine volumes of technical papers presented at the SEM 2017 SEM Annual Conference & Exposition on Experimental and Applied Mechanics organized by the Society for Experimental Mechanics and held in Indianapolis, IN, June 12–15, 2017. The complete Proceedings also includes volumes on: *Dynamic Behavior of Materials; Challenges In Mechanics of Time-Dependent Materials; Mechanics of Biological Systems, Materials and Other Topics in Experimental and Applied Mechanics; Micro-and Nanomechanics; Mechanics of Composite, Hybrid & Multifunctional Materials; Fracture, Fatigue, Failure and Damage Evolution; Residual Stress, Thermomechanics & Infrared Imaging, Hybrid Techniques and Inverse Problems; and Mechanics of Additive and Advanced Manufacturing.*

Each collection presents early findings from experimental and computational investigations on an important area within experimental mechanics, optical methods being one of these areas.

With the advancement in imaging instrumentation, lighting resources, computational power, and data storage, optical methods have gained wide applications across the experimental mechanics society during the past decades. These methods have been applied for measurements over a wide range of spatial domain and temporal resolution. Optical methods have utilized a full range of wavelengths from X-ray to visible lights and infrared. They have been developed not only to make two-dimensional and three-dimensional deformation measurements on the surface, but also to make volumetric measurements throughout the interior of a material body.

Bari, Italy — Luciano Lamberti
Taichung, Taiwan — Ming-Tzer Lin
Worcester, MA, USA — Cosme Furlong
Chicago, IL, USA — Cesar Sciammarella

# Contents


1 **A New Method for Improving Measurement Accuracy of Digital Image Correlation** ... 1
Li Bang-Jian, Wang Quan-Bao, and Duan Deng-Ping

2 **Fatigue Analysis of 7075 Aluminum Alloy by Optoacoustic Method** ... 7
Tomohiro Sasaki, Hiroshi Ono, Sanichiro Yoshida, and Shuich Sakamoto

3 **Early Strain Localization in Strong Work Hardening Aluminum Alloy (2198 T3): 3D Laminography and DVC Measurement** ... 15
Ante Buljac, Lukas Helfen, François Hild, and Thilo F. Morgeneyer

4 **On the In-Plane Displacement Measurement by 3D Digital Image Correlation Method** ... 19
Chi-Hung Hwang, Shou Hsueh Wang, and Wei-Chung Wang

5 **Noise Reduction in Amplitude-Fluctuation Electronic Speckle-Pattern Interferometry** ... 27
Sanichiro Yoshida, David Didie, Jong-Sung Kim, and Ik-Keun Park

6 **Evaluating Path of Stress Triaxiality to Fracture of Thin Steel Sheet Using Stereovision** ... 37
D. Kanazawa, S. Chinzei, Y. Zhang, K. Ushijima, J. Naito, and S. Yoneyama

7 **Studying with a Full-Field Measurement Technique the Local Response of Asphalt Specimens Subjected to Freeze-Thaw Cycles** ... 43
M.C. Teguedi, B. Blaysat, E. Toussaint, S. Moreira, S. Liandrat, and M. Grédiac

8 **Mechanical Shape Correlation: A Novel Integrated Digital Image Correlation Approach** ... 47
S.M. Kleinendorst, J.P.M. Hoefnagels, and M.G.D. Geers

9 **On the Boundary Conditions and Optimization Methods in Integrated Digital Image Correlation** ... 55
S.M. Kleinendorst, B.J. Verhaegh, J.P.M. Hoefnagels, A. Ruybalid, O. van der Sluis, and M.G.D. Geers

10 **Extension of the Monogenic Phasor Method to Extract Displacements and Their Derivatives from 3-D Fringe Patterns** ... 63
C.A. Sciammarella and L. Lamberti

11 **Deformation Measurement within a Volume of Translucent Yield Stress Material Using Digital Image Correlation** ... 77
A. McGhee and P. Ifju

12 **Surface Deformation with Simultaneous Contact Area Measurement for Soft Transparent Media due to Spherical Contact** ... 81
A. McGhee, D. Nguyen, and P. Ifju

13 **Towards Measuring Intergranular Force Transmission Using Confocal Microscopy and Digital Volume Correlation** ... 85
Kimberley Mac Donald and Guruswami Ravichandran

14 **Using Anti-aliasing Camera Filters for DIC: Does It Make a Difference?** ... 89
PL. Reu

**15 Investigation of Electronic Speckle Pattern Interferometry with Line Laser Scanning for Large Area Deformation Measurement** ........ 93
Shuichi Arikawa and Yuta Ando

**16 Internal Heat Generation in Tension Tests of AISI 316 Using Full-Field Temperature and Strain Measurements** ........ 97
Jarrod L. Smith, Veli-Tapani Kuokkala, Jeremy D. Seidt, and Amos Gilat

**17 A Short Survey on Residual Stress Measurements by HDM and ESPI** ........ 105
C. Pappalettere

**18 Feasibility of Using Fringe Projection System for Corrosion Monitoring in Metals of Interest in Cultural Heritage** ........ 111
C. Casavola, P. Pappalardi, G. Pappalettera, and G. Renna

# Chapter 1
# A New Method for Improving Measurement Accuracy of Digital Image Correlation

**Li Bang-Jian, Wang Quan-Bao, and Duan Deng-Ping**

**Abstract** Digital image correlation (DIC) method has been applied in wide fields including experimental mechanics. The displacement measurement accuracy plays an important role in these situations. The direct method to improve the measurement accuracy is to reduce the measurement error in DIC. In the paper, a new method has been developed to improve the DIC accuracy using the feature of DIC system error. The feature of DIC system error is analyzed. And the reduction of the DIC system error has been verified by experiment.

**Keywords** Digital image correlation • System error • Translation measurement • Error reduction • Measurement accuracy

## 1.1 Introduction

Digital image correlation (DIC) is an optical measurement method and DIC is popular in motion measurement, dimension measurement and experimental mechanics for its countless and full-field features [1–5]. In 2D DIC, a digital camera is used to capture the reference image before the specimen deformation and capture the deformed image when the loading is imposed by drawing machine. In motion measurement, the reference and deformed images are captured before and after the specimen or object movement respectively. In dimension measurement, the same speckle images in different positions are regarded as the reference and deformed images respectively. And in subset based DIC, the displacement of sampled point can be obtained by matching the selected subset. Subsequently the dynamic variables including velocity and acceleration can be calculated by post-processing with the displacement data. That is to say that displacement or translation measurement is the basic for other variables' calculation. So the translation measurement of DIC has been studied by many researchers [6–8]. And the measurement accuracy of DIC has been one main research point recently [9, 10]. In addition, the DIC calculation speed has been another main research point.

For calculation speed, many techniques are proposed. Pan [11] proposed that reliability guided searching method combined with interpolation coefficient pre-computed technique can make the calculation speed about 160 times faster than the classical DIC method. Jiang *et al.* [12] presented that path-independent searching method combined with graphics processing unit (GPU) based parallel calculation, inverse compositional Gauss Newton (IC-GN) [13] and Fast Fourier transform (FFT) can make about 66 times faster than non-path-independent method.

For measurement accuracy, many researchers have studied it by theory, simulation and experiments. Generally measurement accuracy can be quantified with measurement error which contains system error and random error by theoretical and experimental analysis. For instance, Wang *et al.* [14] investigated the shape function induced random error theoretically and experimentally. Schreier *et al.* [15] studied the interpolation induced system error by simulation. Pan *et al.* [8] investigated lens distortion induced system error by theory and experiments. Ma *et al.* [7] researched the camera self-heating induced system error by experiments. Su and Xu *et al.* [16–18] derived the system error theoretically. Hu *et al.* [19] developed a method to evaluate 3D DIC error. Pan *et al.* [20, 21] used bilateral telecentric lens to improve the 2D DIC measurement accuracy. Considering the limitation of field of view of bilateral telecentric lens, Pan *et al.* [22] used a non-deformable sample to compensate the DIC measurement accuracy. Pan [23] and Zhou *et al.* [24] proposed that image pre-filtering in DIC can enhance measurement accuracy by smoothing the image noise but the pre-filtering parameters had to be selected and the random error increased due to the pre-filtering smoothing function. Zhou *et al.* [25] developed an adaptive subset offset method in incremental DIC to improve the measurement accuracy by avoiding interpolation of the sample points.

---

L. Bang-Jian • W. Quan-Bao (✉) • D. Deng-Ping
School of Aeronautics and Astronautics, Shanghai Jiao Tong University, Shanghai, People's Republic of China
e-mail: quanbaowang@sjtu.edu.cn



L. Lamberti et al. (eds.), *Advancement of Optical Methods in Experimental Mechanics, Volume 3*,
Conference Proceedings of the Society for Experimental Mechanics Series, DOI 10.1007/978-3-319-63028-1_1

In this paper, a novel method to reduce error will be proposed which is different from existing solutions which focus on random error and based on image pre-filtering. The new strategy is developed from the system error reduction perspective to improve the measurement accuracy. Due to the fact that systematic error is related with interpolation, noise, shape function, correlation criterion and so on, a few effective methods to reduce the systematic error are studied. In this paper, a strategy will be put forward based on the shape feature of systematic errors in translation measurement using DIC.

To sum up, the improvement of the translation or displacement measurement accuracy plays an important role in practical applications, especially in motion and dimension measurement. The quasi odd function feature of the system error in DIC will be exhibited by simulation experiments. Subsequently the strategy will be proposed based on the quasi odd function feature. And the system error reduction will be derived mathematically in detail. Finally, the open access speckle image data will be applied to verify the effectiveness of the developed strategy.

## 1.2 Algorithm

### *1.2.1 Quasi Odd Function Feature of DIC System Error*

In this part, the open access speckle images from DIC Challenge Datasets (https://sem.org/dic-challenge/) are applied to implement the experiment which can show the quasi odd function feature of the system error. In detail, the reference speckle image is selected from the DIC Challenge 2D Datasets. Subsequently by translating, one deformed speckle image set is built. The Fourier Shift theorem [1, 15] is used to translate the reference speckle image to form the deformed set. The displacements are from $-0.5$ to 0.5 pixels and the displacement interval is 0.1 pixels. The subset based local DIC algorithm with first order shape function and bi-cubic interpolation is used to calculate the displacements. And the subset size is set as $25 \times 25$ pixels and the area of interest (AOI) is $100 \times 100$ pixels. And the mean bias error is used to quantify the system error. In addition, $2 \times 2$, $4 \times 4$, $6 \times 6$, $8 \times 8$ and $10 \times 10$ kernel based bi-cubic interpolations are used to demonstrate the quasi odd function feature too. As showed in Fig. 1.1, the quasi odd function feature of DIC system error for different interpolations is evident.

### *1.2.2 Strategy*

On the basis of the quasi odd function feature of DIC system error, a system error reduction strategy is developed. As displayed in Fig. 1.1, the system errors of the symmetric displacements are opposite numbers to each other. So the system error of the displacement can be reduced or removed by the system error of the symmetric displacement. In the proposed

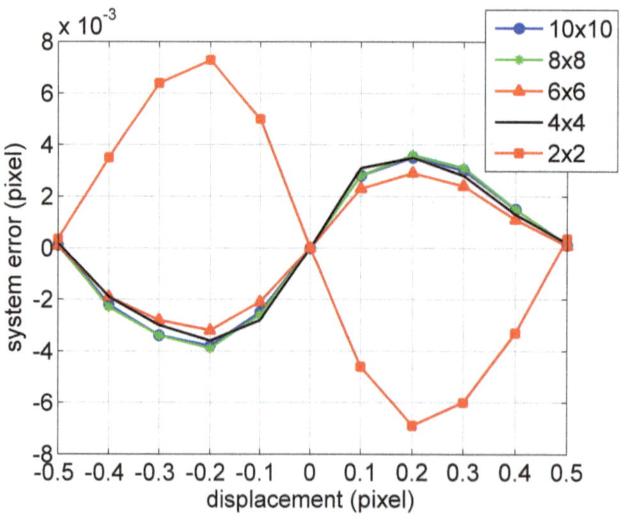


**Fig. 1.1** The quasi odd function feature of system error induced by different bi-cubic interpolation methods in digital image correlation (DIC)

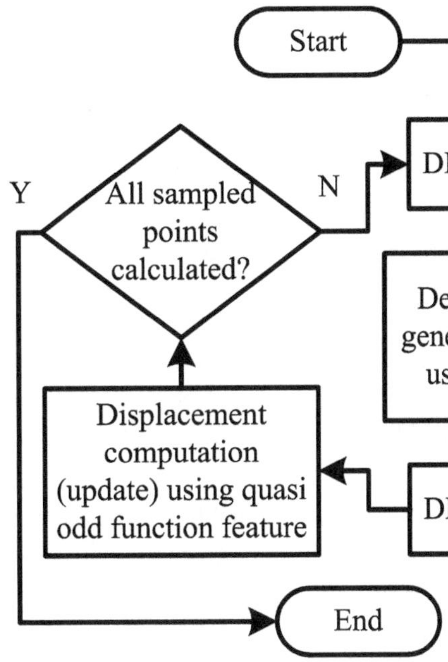


**Fig. 1.2** Schematic of system error reduction strategy using quasi odd function feature

strategy, the deformed speckle image with the symmetric displacement is obtained by Fourier Shift theorem. Due to the fact that the true displacement of the sampled point is unknown in practical application, the symmetric displacement is estimated by the traditional DIC calculation. Here it should be noted that the updated deformed speckle image with the symmetric displacement is generated from the original deformed image (not the reference image), since the noises of the original deformed and reference images are different. So the amplitude of the final translated displacement is double of that of the symmetric displacement estimated by DIC calculation.

As displayed in Fig. 1.2, the first step of the developed strategy is implementing the traditional DIC calculation between the original reference image and original deformed image. The calculated displacement is assumed as $u_c$. Subsequently, the updated deformed image is generated by translating the original deformed image with -$2u_c$ displacement. Then DIC calculation is carried out again between the original reference image and the updated deformed image. And the second calculated displacement is assumed as $u_{c2}$. Next the updated displacement $u_u$ can be obtained by the following formula:

$$u_u = \frac{u_c + u_{c2} - (-2u_c)}{2} = \frac{3u_c + u_{c2}}{2}. \tag{1.1}$$

The sampled points are judged whether calculated or not in the final step of the proposed strategy.

Mathematically, the system error of the proposed strategy can be derived. The true displacement of the sampled point is assumed as $u$. and the corresponding system error is denoted by $e(u)$. As displayed in Fig. 1.1, the quasi odd function feature can be written as

$$-e(u) = e(-u). \tag{1.2}$$

In addition, the first calculated displacement $u_c$ can be written as

$$u_c = u + e(u). \tag{1.3}$$

Similarly, the second calculated displacement $u_{c2}$ can be written as

$$u_{c2} = u - 2u_c + e(u - 2u_c). \tag{1.4}$$

Then the updated displacement $u_u$ can be derived by substituting formulas (1.3) and (1.4) into (1.1):

$$u_u = \frac{2u + e(u) + e\left[-u - 2e(u)\right]}{2}. \tag{1.5}$$

According to the quasi odd function feature of the system error and formula (1.2), the updated displacement $u_u$ and formula (1.5) can be written as:

$$u_u = u + \frac{e(u) - e\left[u + 2e(u)\right]}{2}. \tag{1.6}$$

From the above derivation, the system error of the updated displacement can be written as:

$$e\left(u_u\right) = \frac{e(u) - e\left[u + 2e(u)\right]}{2}. \tag{1.7}$$

When $e(u)$ is small and generally this situation is tenable with suitable interpolation method, $u + 2e(u)$ will be close to $u$ and then $e(u_u)$ will be smaller than $e(u)$. That is to say, the system error will be reduced by the proposed strategy mathematically.

## 1.3 Experiments

The open access speckle images (https://sem.org/dic-challenge/) will be used to verify the proposed strategy too. Measurement accuracy will be investigated in this part.

The proposed strategy is tested with the open access speckle images (https://sem.org/dic-challenge/). The open access speckle image is selected as reference image. And one deformed image set is obtained by translating. The deformed image set contains 11 deformed images which are generated by Fourier shift theorem with displacement from −0.5 to 0.5 pixels and the displacement interval is 0.1 pixels. In the experiments, the area of interest (AOI) is 101 × 101 pixels. The subset size is 25 × 25 pixels. The subset based local DIC with first order shape function and bi-cubic interpolation is utilized. The system error which is calculated by the mean bias error is displayed in Fig. 1.3. As showed in Fig. 1.3, the system error is reduced for different bi-cubic interpolations by the proposed strategy. In addition, the result of 2 × 2 bi-cubic interpolation is worse than others, which can be explained that 2 × 2 bi-cubic interpolation is similar to bi-linear interpolation which is low order interpolation method [26]. Here it should be noted that the reduced system error is not equal to zero for many other factors induced system error such as shape function [27] and correlation criterion [28].

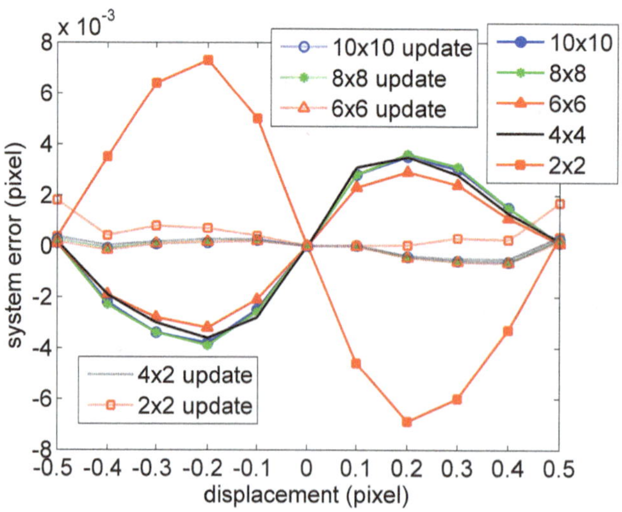


**Fig. 1.3** System error reduction for different bi-cubic interpolation methods (update denotes the proposed strategy)

## 1.4 Conclusion

The quasi odd function feature of system error in DIC is displayed for different interpolations. An effective system error reduction strategy is proposed using the quasi odd function feature. Open access speckle images (https://sem.org/dic-challenge/) are used to test the proposed strategy. The measurement accuracy of the proposed strategy is investigated. And the results indicate that the system error can be reduced by the proposed strategy.

In addition, the developed strategy is suitable for the situations including motion measurement and dimension measurement which can employ the DIC technique. And the simulated speckle pattern can be pasted on the moving target or the same speckle patterns can be pasted on different positions of the specimen whose dimension needs to be measured.

## References


1. Sutton, M.A., Orteu, J.J., Schreier, H.: Image Correlation for Shape, Motion and Deformation Measurements: Basic Concepts, Theory and Applications. Springer Science & Business Media, New York (2009)
2. Pan, B., et al.: Two-dimensional digital image correlation for in-plane displacement and strain measurement: a review. Meas. Sci. Technol. **20**(6), 062001 (2009)
3. Wu, R., Qian, H., Zhang, D.: Robust full-field measurement considering rotation using digital image correlation. Meas. Sci. Technol. **27**(10), 105002 (2016)
4. Ashrafi, M., Tuttle, M.E.: Measurement of strain gradients using digital image correlation by applying printed-speckle patterns. Exp. Tech. **40**(2), 1–7 (2016)
5. Zhao, J., Zhao, D., Zhang, Z.: A non-contact varying temperature strain measuring system based on digital image correlation. Exp. Tech. **40**(1), 101–110 (2016)
6. Wang, D., et al.: Bias reduction in sub-pixel image registration based on the anti-symmetric feature. Meas. Sci. Technol. **27**(3), 035206 (2016)
7. Shaopeng, M.A., Pang, J., Qinwei, M.A.: The systematic error in digital image correlation induced by self-heating of a digital camera. Meas. Sci. Technol. **23**(2), 344–347 (2012)
8. Pan, B., et al.: Systematic errors in two-dimensional digital image correlation due to lens distortion. Opt. Lasers Eng. **51**(2), 140–147 (2013)
9. Wang, Y., et al., Theoretical analysis on the measurement errors of local 2D DIC: part II assessment of strain errors of the local smoothing method–approaching an answer to the overlap question. Strain. 52(2):129–147, 2016
10. Li, B., Wang, Q., Duan, D.: A modified digital image correlation with enhanced speed and improved accuracy. Proc. SPIE. (2017)
11. Pan, B.: Reliability-guided digital image correlation for image deformation measurement. Appl. Opt. **48**(8), 1535–1542 (2009)
12. Jiang, Z., et al.: Path-independent digital image correlation with high accuracy, speed and robustness. Opt. Lasers Eng. **65**, 93–102 (2015)
13. Shao, X., et al.: Real-time 3D digital image correlation method and its application in human pulse monitoring. Appl. Opt. **55**(4), 696–704 (2016)
14. Wang, B., Pan, B.: Random errors in digital image correlation due to matched or overmatched shape functions. Exp. Mech. **55**(9), 1–11 (2015)
15. Schreier, H.W., Braasch, J.R., Sutton, M.A.: Systematic errors in digital image correlation caused by intensity interpolation. Opt. Eng. **39**(11), 2915–2921 (2000)
16. Su, Y., et al.: Fourier-based interpolation bias prediction in digital image correlation. Opt. Express. **23**(15), 19242–19260 (2015)
17. Su, Y., et al.: Noise-induced bias for convolution-based interpolation in digital image correlation. Opt. Express. **24**(2), 1175 (2016)
18. Xu, X., Su, Y., Zhang, Q.: Theoretical estimation of systematic errors in local deformation measurements using digital image correlation. Opt. Lasers. Eng. **88**, 265–279 (2017)
19. Hu, Z., et al.: Error evaluation technique for three-dimensional digital image correlation. Appl. Opt. **50**(33), 6239–6247 (2011)
20. Pan, B., Yu, L., Wu, D.: High-accuracy 2D digital image correlation measurements with bilateral telecentric lenses: error analysis and experimental verification. Exp. Mech. **53**(9), 1719–1733 (2013)
21. Pan, B., Yu, L., Wu, D.: Accurate ex situ deformation measurement using an ultra-stable two-dimensional digital image correlation system. Appl. Opt. **53**(19), 4216–4227 (2014)
22. Pan, B., Yu, L., Wu, D.: High-accuracy 2D digital image correlation measurements using low-cost imaging lenses: implementation of a generalized compensation method. Meas. Sci. Technol. **25**(2), 025001–025012 (2014)
23. Pan, B.: Bias error reduction of digital image correlation using Gaussian pre-filtering. Opt. Lasers. Eng. **51**(10), 1161–1167 (2013)
24. Zhou, Y., et al.: Image pre-filtering for measurement error reduction in digital image correlation. Opt. Lasers. Eng. **65**(1), 46–56 (2015)
25. Zhou, Y., Sun, C., Chen, J.: Adaptive subset offset for systematic error reduction in incremental digital image correlation. Opt. Lasers. Eng. **55**(7), 5–11 (2014)
26. Luu, L., et al.: Accuracy enhancement of digital image correlation with B-spline interpolation. Opt. Lett. **36**(16), 3070–3072 (2011)
27. Yu, L., Pan, B.: The errors in digital image correlation due to overmatched shape functions. Meas. Sci. Technol. **26**(4), 045202 (2015)
28. Bing, P., et al.: Performance of sub-pixel registration algorithms in digital image correlation. Meas. Sci. Technol. **17**(6), 1615 (2006)


**Li Bang-Jian** is a PhD candidate in the School of Aeronautics and Astronautics at Shanghai Jiao Tong University. His research interests include digital image correlation and its applications in control science and engineering.

# Chapter 2
# Fatigue Analysis of 7075 Aluminum Alloy by Optoacoustic Method

**Tomohiro Sasaki, Hiroshi Ono, Sanichiro Yoshida, and Shuich Sakamoto**

**Abstract** The influence of fatigue damage on the elastic response of AA7075 aluminum alloy was investigated through a combination of optical and acoustical experiments. Specimens were previously subjected to fatigue cyclic loads at various fatigue levels within the fatigue life. Macroscopic deformation process under a certain load below the yield point (elastic region) for the pre-fatigued specimen was visualized by electronic speckle pattern interferometry (ESPI). At the same time, the acoustic velocities of vertical and shear waves propagating in the fatigued specimen were measured using an ultrasonic probe. The acoustic analysis showed the following change in residual stress by the fatigue cyclic load; an increase in compressive residual stress with the number of pre-fatigue cycles ($N_P$) below $10^3$, and relaxation of the residual stress $N_P$ over $10^3$. The visualization using ESPI demonstrated that the strain heterogeneity in the macroscopic elastic regime was enhanced with increase of the pre-fatigue cycle. The correlation between the optical and the acoustical measurement results is discussed based on the change in the residual stress, localized plastic deformation, and the crack initiation.

**Keywords** Fatigue • Optical method • Acoustic method • Speckle patter interferometry • Aluminum alloy

## 2.1 Introduction

Fatigue of metals is generally interpreted as a process of crack initiation by localized deformation, and crack propagation, leading to final fracture under cyclic loading. A number of fatigue inspection methods such as those use X-ray, ultrasonic wave, and acoustic emission have been established [1–4]. These methods mainly aim at detection of the presence of fatigue crack, and fatigue life is predicted by monitoring the crack length based on fracture mechanical parameters. On the other hand, the stage of localized plastic deformation prior to the crack initiation is a complex phenomenon and not fully understood. Thus, it is generally difficult to detect the fatigue damage at the earlier fatigue stage particularly in ductile metals, because most of the fatigue life is spent by the stage before the crack initiation. The localized plastic deformation occurs not only in "Low cycle fatigue", but also in "High cycle fatigue" that is characterized by lower stress condition below the macroscopic yield stress. This study focuses on the influence of the localized plastic deformation on macroscopic deformation behavior of metals. Our previous works using a full field optical method [5, 6] demonstrated that the strain concentration, which is observed in the macroscopic elastic regime, is enhanced depending on the degree of fatigue cycles. On the other hand, a similar effect of fatigue damage on the elastic behavior of metals was also confirmed through acoustical methods by several researchers [7–10]. These methods are based on stress dependence of elastic wave velocity that propagates through the material, termed as "acousto-elasticity". The fatigue damage was detected as a change in the elastic wave velocity or attenuation attributed to residual stress induced by the localized plastic deformation. Both the results obtained from the optical and the acoustical methods are associated with the elastic response of metals, and indicate the possibility of analyzing the elastic response as a mean of fatigue diagnosis.

T. Sasaki (✉) • H. Ono • S. Sakamoto
Graduate school of Niigata University, 8050 Ikarashi-ninocho, Nishi-ku, Niigata-shi, Niigata, Japan
e-mail: tomodx@eng.niigata-u.ac.jp

S. Yoshida
Department of Chemistry and Physics, Southeastern Louisiana University, SLU 10878, Hammond, LA, 70402, USA
e-mail: syoshida@selu.edu



L. Lamberti et al. (eds.), *Advancement of Optical Methods in Experimental Mechanics, Volume 3*,
Conference Proceedings of the Society for Experimental Mechanics Series, DOI 10.1007/978-3-319-63028-1_2

In the present study, in order to establish the correlation of the elasticity of metals and the fatigue damage, the deformation behavior of aluminum alloys previously subjected to cyclic loading are visualized with electronic speckle pattern interferometry (ESPI). At the same time, the change of elasticity due to the fatigue damage is evaluated by the acoustic measurement. We demonstrate the influence of microscopic factors including the localized plastic deformation, the presence of micro-cracks, on strain heterogeneity, and discuss the change in elastic response of the material.

## 2.2 Experimental Procedure

### 2.2.1 *Fatigue Test*

An AA7075-T1 aluminum alloy sheet with thickness of 3.0 mm was used for experiments. Necked shape specimens as shown in Fig. 2.1 were cut from the sheet for fatigue tests by electronic discharge machining. The yield strength of specimen measured by a static tensile test was approximately 8.0 kN. The fatigue test was conducted for the specimen under load control condition with a sinusoidal waveform of 5 Hz. The maximum tensile load ranged from 5.0 kN to 7.0 kN, and the minimum load was a constant at 0.1kN. Figure 2.2 shows the maximum load – number of cycle curve obtained from the fatigue test. From this curve, the maximum load in the pre-fatigue test was determined to 5.0 kN which was approximately the fatigue limit. The numbers of cycles used in the pre-fatigue test, $N_P$, were 10, $10^2$, $10^3$, $10^4$, and $10^5$, respectively.

### 2.2.2 *Acoustic Measurement*

Acoustic wave velocity in the specimen was measured using an ultrasonic transducer driven by a square wave pulser-reciever with 35 MHz bandwidth (Olympus 5077PR). As shown in Fig. 2.3, the measurement was conducted for the necked center of specimen. Velocities of vertical wave propagating in the thickness direction ($V_{zz}$), and shear vertical waves in the displacement directions of $x$ and $y$ ($V_{zx}$, $V_{zy}$) were respectively measured. Coupling media used was distilled water for the vertical wave measurement, and glycerin paste for the shear wave measurement.

### 2.2.3 *Dynamic Observation of Deformation Behavior of Pre-fatigued specimen*

Deformation behavior during tensile tests was visualized with a two-dimensional ESPI in the same manner described in a previous work [6]. Figure 2.4 shows the optical setup in this study. Two optical interferometers were arranged to horizontal

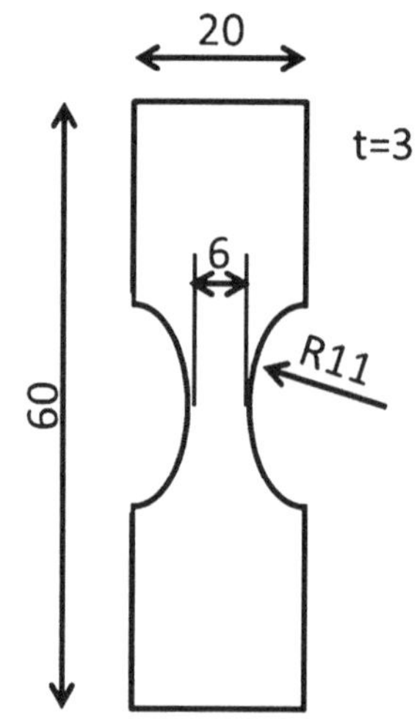


**Fig. 2.1** Specimen used in this study

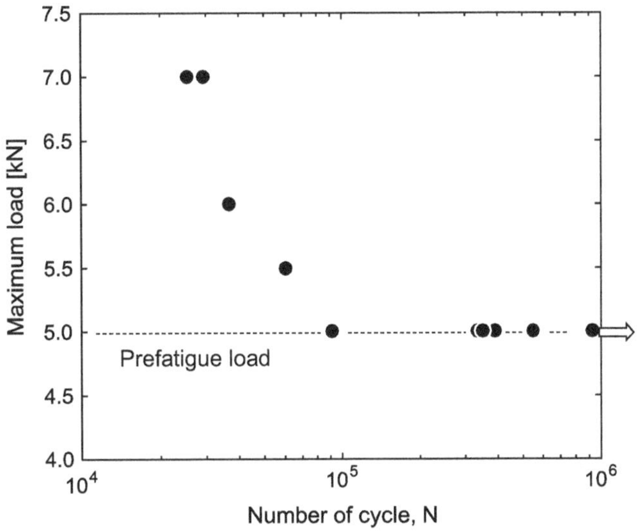


**Fig. 2.2** Maximum load vs. number of cycle plot in the fatigue test

**Fig. 2.3** Acoustic measurement

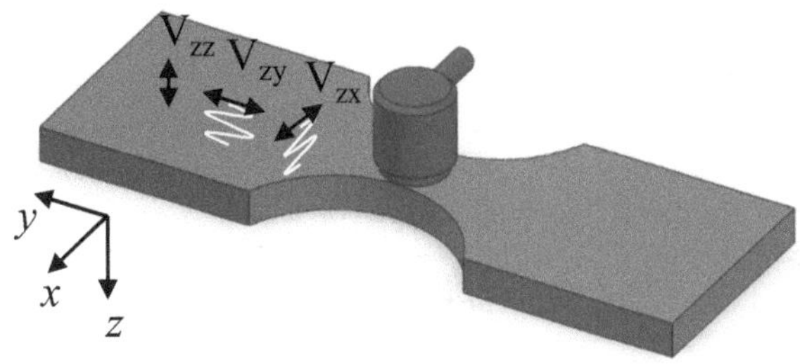


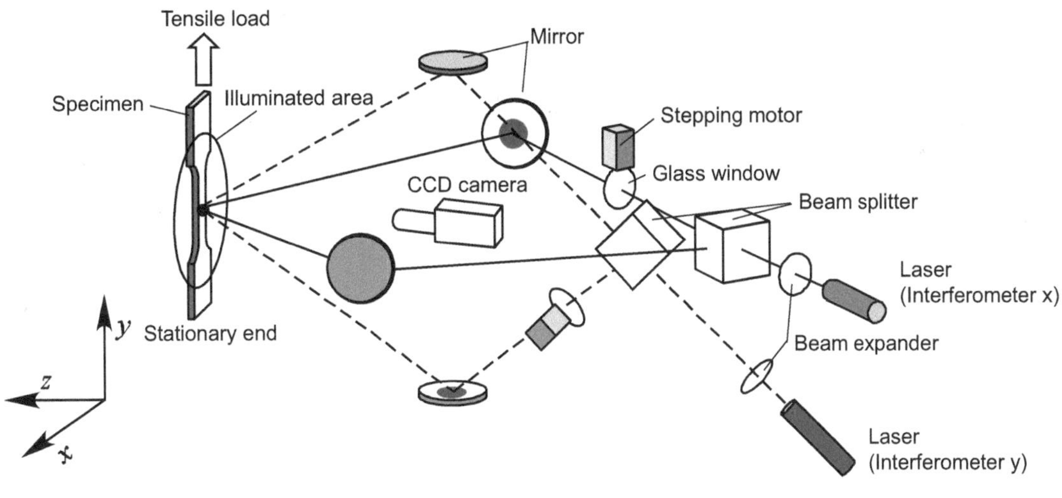


**Fig. 2.4** Experimental arrangement of 2D-ESPI [6]

direction ($x$) and vertical direction ($y$) to the tensile machine (interferometer $x$, and interferometer $y$). Each interferometer was sensitive to in-plane displacement along its horizontal direction. Semiconductor laser with the wave length of 660 nm and the power of 50 mW was used for light source. The individual displacement fields in each sensitive direction were obtained by switching light sources of the two interferometers. The laser beam was expanded by a beam expander and split into two paths by a beam splitter and converged to the surface of specimen via two mirrors. The incident angels to the surface were 36.3° for the interferometer $x$, and 48.0° for the interferometer $y$. The speckle pattern was captured with a CCD camera with a frame rate of 15 frames per a second. The speckle intensity received by each pixel changes depending on the displacement in the sensitive direction due to optical path difference between the two interferometric arms. Thus, the displacement field on the measured surface can be obtained by computing the intensity difference as fringe contours.

In addition, a glass window with 5.2 mm thick and the refraction index of 1.53 was placed after the beam splitter for one interferometric arm of each interferometer, in order to introduce "a carrier fringe system" [11]. The optical distance of the laser beam which passes through the glass window varies depending on its incident angle. Since the beam is expanded by the expander, optical distance on the irradiated surface has a gradient. By rotating the window with a stepping motor, the carrier fringes orthogonal to the sensitive direction appear. The resultant fringe contours represents the superimposing displacement obtained from the carrier and the displacement, thus actual displacement can be obtained by subtracting the carrier from the measured fringe. The number of carrier fringes introduced to the measurement surface was 6 for the horizontal direction, *x*, and 3 for the tensile direction, *y*. Tensile load was applied up to 800 N (5% of the yield load), and the dynamic deformation behavior was measured.

## 2.3 Result and Discussion

### *2.3.1 Effect of Fatigue Cycle on Acoustic Wave Velocity*

Figure 2.5 shows acoustic velocity plotted against logarithm of the number pre-fatigued cycles. The value of non-fatigued specimen ($N_P = 0$) pis plotted to the pre-fatigue cycles, $N_P = 10^0$ for convenience. The vertical wave propagating in the thickness direction, $V_{zz}$, decreased as $N_P$ increased up to $10^3$, then it increased again (Fig. 2.5a). In acousto-elastic theory, the elastic modulus depends on applied stress due to the non-linearity between the interatomic force and the interatomic distance [11]. In particular, tensile stress leads to a decrease in the sound velocity. Thus, the initial decrease in the $V_{zz}$ at $N_P < 10^3$ is indicative of an increase of the tensile internal stress in the thickness direction, *z*. On the other hand, in the results of shear waves (Fig. 2.5b), although the change was relatively small, both $V_{zx}$ and $V_{zy}$ exhibited a slight increase at lower $N_P$, followed by a decrease in contrast to the change of $V_{zz}$. Toda et al. [12] proposed "R-value acoustoelastic method" that uses the ratio of vertical wave velocity and averaged value of shear wave velocity. According to this method, the ratio of $V_{zz}$. and $(V_{zx} + V_{zy})/2$ is proportional to the sum of in-plane principle stress as shown below.

$$R = \frac{Vzz}{\left(V_{zx} + V_{zy}\right)/2} = R_0 + C_R\left(\sigma_x + \sigma_y\right) \tag{2.1}$$

where $R_0$ is microstructural factor, $C_R$ is stress-acoustic constant, $\sigma_x$, and $\sigma_y$ are components of plane stress. Figure 2.6 shows the *R* value plotted against the pre-fatigue cycle. Since the R-value generally has a positive value, a decrease at

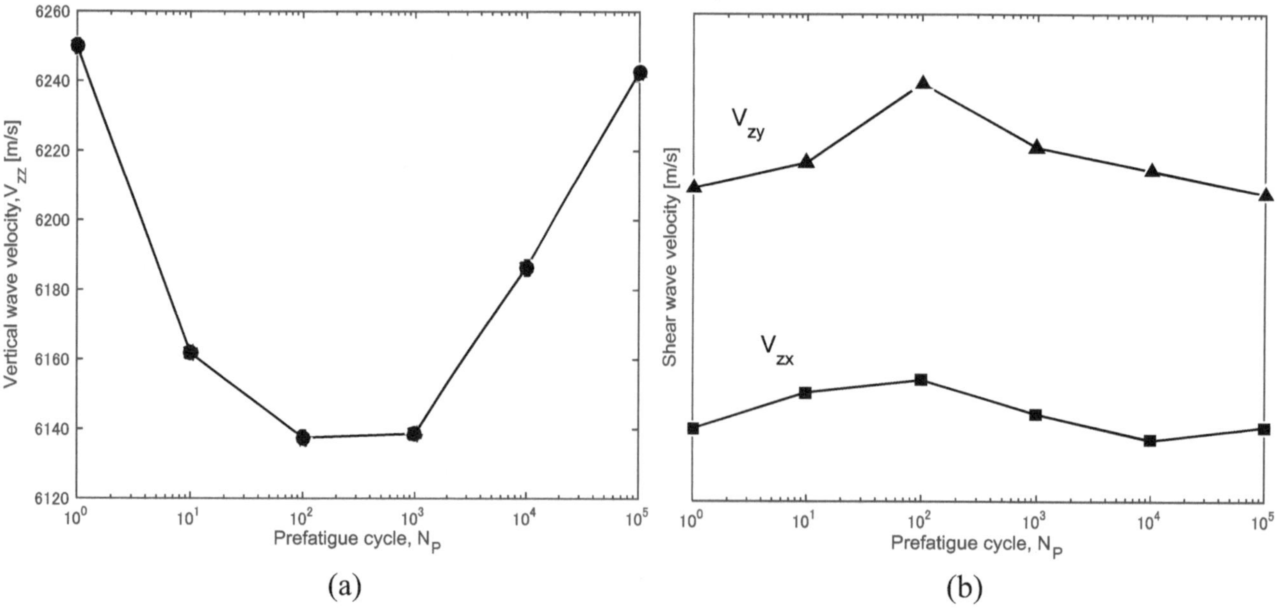


**Fig. 2.5** Acoustic wave velocity for (**a**) vertical wave $V_{zz}$, and (**b**) shear vertical waves $V_{zx}$ and $V_{zy}$

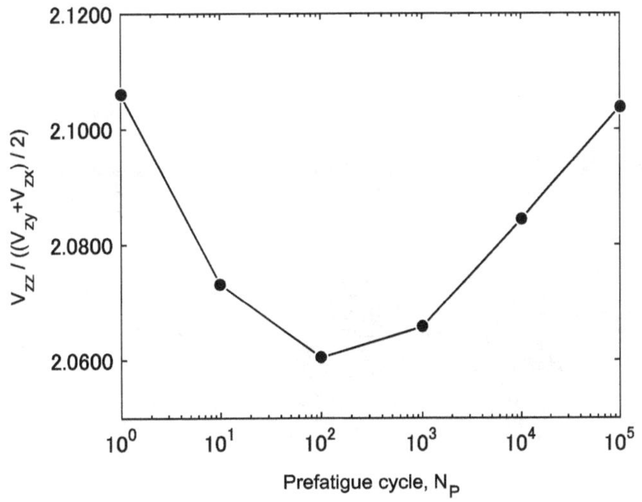


**Fig. 2.6** The ratio of vertical and averaged shear wave velocity

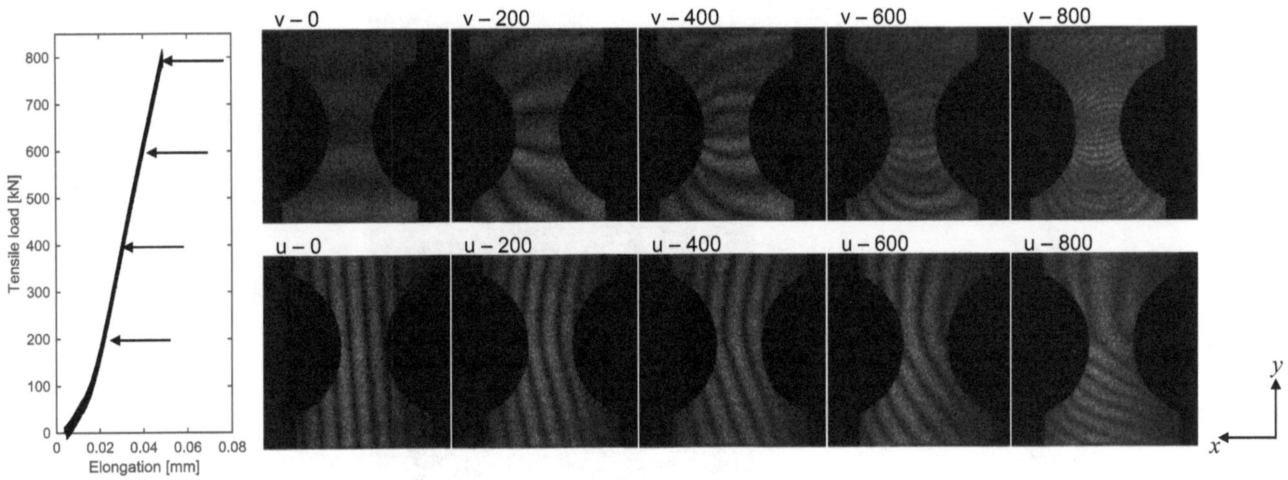


**Fig. 2.7** Load-elongation curve and fringe pattern observed at various tensile loads of a non-fatigued specimen. The tensile load is indicated in the load – elongation curve by *arrows*

$N_P < 10^3$ implies an increase of compressive stress in the x-y plane. Cyclic tensile load during the pre-fatigue causes the localized plastic deformation in the necked area. This fact might result in the increase of compressive residual stress, because the deformed area is restrained by the surrounding area. Therefore, the earlier pre-fatigue stage of $N_P < 10^3$ can be interpreted as an increasing process of the residual stress in the necked part. Furthermore, the latter increase in the R-value at $N_P > 10^3$ indicates a relaxation of the residual stress. This relaxation behavior agrees with the result of surface acoustic wave velocity measurement by a scanning acoustic microscope in our previous work [6]. In the previous work, crack initiation was confirmed at the latter fatigue stage when the acoustic wave velocity decreased. It can be understood that the residual stress induced by the localized plastic deformation was relaxed with the crack initiation.

### *2.3.2 Deformation Behavior in Tensile Test*

Figure 2.7 shows a load-elongation curve and optical fringe patterns observed in a tensile test of the non-fatigued specimen ($N_P = 0$). These fringe patterns represent displacement contours along *x*, and *y* resulting from subtracting captured images at various tensile load levels by a base image before the tensile test (tensile load = 0 N). Fringe pattern at 0 N (*v*-0 N or *u*-0 N) means the carrier fringes initially introduced. The fringe in the tensile direction *y* (*v*-fringe) concentrated to the neck

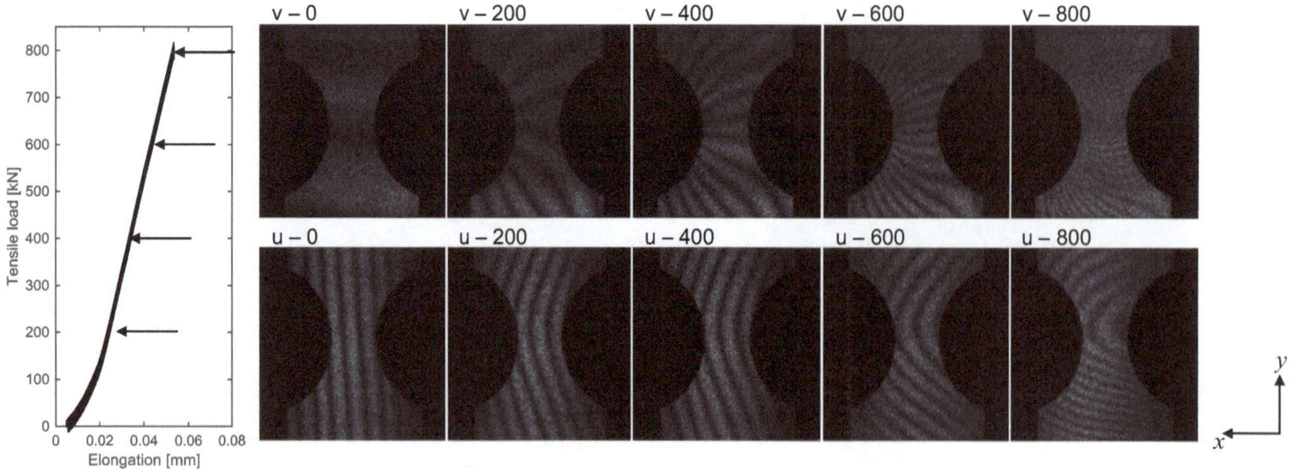


**Fig. 2.8** Load-displacement curve and fringe pattern observed of a pre-fatigued specimen of $N_P = 10^5$

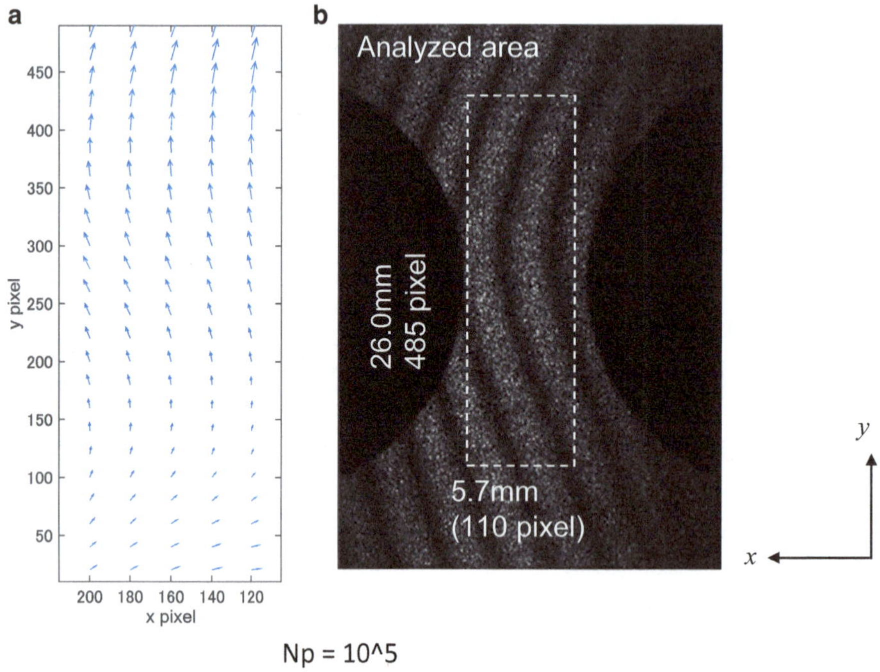


**Fig. 2.9** Displacement vector field showing the deformation of necked part obtained from fringe analysis. The analyzed area is indicated in the image on the right

part of specimen with increase of tensile load, showing the strain concentration. In the horizontal direction, $x$ ($u$-fringe), the initial carrier fringes gradually became a bent shape. Difference between the non-fatigue and pre-fatigued specimen was observed in this strain concentration behavior. Figure 2.8 shows an example of the fringe patterns in a tensile test for the pre-fatigued specimen ($N_P = 10^5$). In all the pre-fatigued condition, the pre-fatigued specimens basically exhibited a similar concentration behavior of the $v$-fringe, whereas the curvature of $u$-fringe tended to be higher than that in the non-fatigued specimen. Figure 2.9 shows displacement vector field of the necked part obtained from phase analysis of the $u$, and $v$ fringe contours. The analysis was conducted for an area of 110 × 485 pixel (5.7 mm × 26.0 mm) around the necked part indicated by dashed line. The quiver plot on the left indicates that the strain concentrates at one side of the neck, implying strain heterogeneity.

To discuss this strain concentration behavior in detail, the mean value of strain in the measured area was computed. Normal strain along the tensile axis $\varepsilon_{yy}$, and shear strain $\gamma_{xy}$ were obtained from the displacement components $v$ and $u$ as follows.

$$\varepsilon_{yy} = \frac{\partial v}{\partial y},\ \gamma_{xy} = \frac{\partial u}{\partial y} + \frac{\partial v}{\partial x} \tag{2.2}$$

Figure 2.8 shows the result plotted against the tensile load up to 400 N. The tensile strain, $\varepsilon_{yy}$ monotonically increased with the increases of applied load as shown in Fig. 2.9a. The pre-fatigued specimens exhibited higher slopes than that in the non-fatigued specimen. In addition, a larger difference was seen in the shear strain $\gamma_{xy}$ as shown in Fig. 2.9b. The increase in $\gamma_{xy}$ implies strain heterogeneity that is probably due to the localized plastic deformation during the fatigue test. The slopes of $\varepsilon_{yy}$ and $\gamma_{xy}$, (compliance $d\varepsilon_{yy}/dF$, and $d\gamma_{xy}/dF$) obtained by a least square method are shown in Fig. 2.10. Variation of normal compliance, $d\varepsilon_{yy}/dF$ by the pre-fatigue cycle was consistent with the measurement result of shear velocity along the tensile direction (Fig. 2.5b); $d\varepsilon_{yy}/dF$ showed an increase at $N_P < 10^3$ and a decrease at $N_P > 10^3$. In addition, the shear compliance, $d\gamma_{xy}/dF$ showed a slight increase at $N_P > 10^3$ and a rapid increase at $N_P < 10^3$. These results demonstrate that the

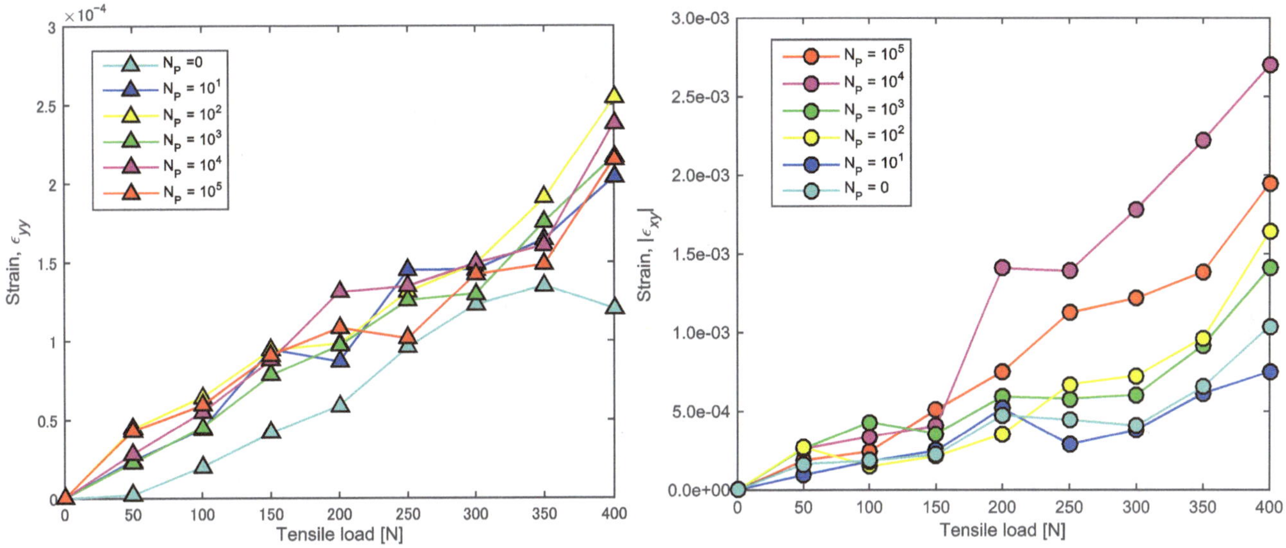


(a) Tensile strain, $\varepsilon_{yy}$, (b) Shear strain, $\gamma_{xy}$

**Fig. 2.10** Mean value of strain plotted against the applied load. (**a**) Tensile strain, $\varepsilon_{yy}$, (**b**) Shear strain, $\gamma_{xy}$

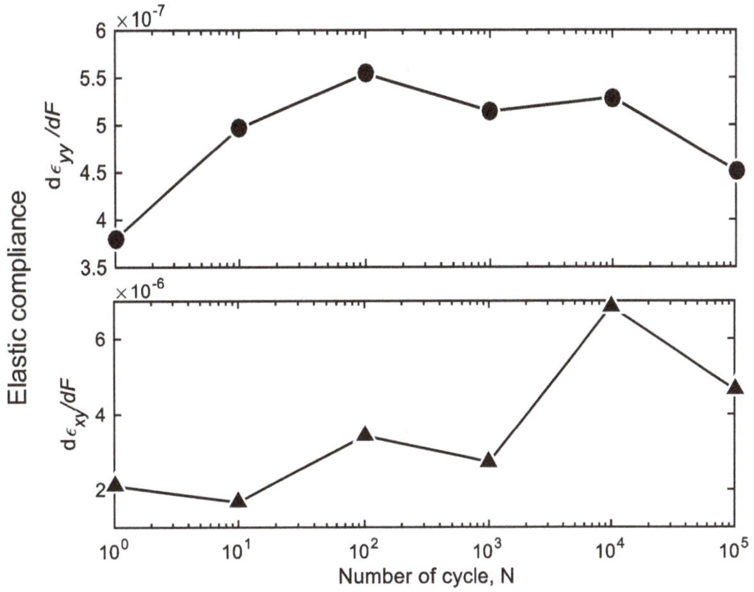


**Fig. 2.11** Elastic compliance

fatigue damage due to the pre-fatigue load influenced elastic deformation behavior as well as the acoustic velocity. However, the acoustic measurement suggests the increase in the elasticity along the tensile direction by compressive residual stress (Fig. 2.6), and this is inconsistent with the increase in the elastic compliance $d\varepsilon_{yy}/dF$. The discrepancy between the acoustic and the optical measurement can be explained by taking into consideration the effect of internal force by residual stress field described in another work [13]. The neck part of specimen is initially compressed by the surrounding region including both grip of specimen. When the specimen is stretched, the restoring force acts on the compressed region, resulting in decreasing the external tensile force. Consequently, the value $d\varepsilon_{yy}/dF$ in the compressed region decreases. At the same time, the heterogeneity of residual stress in both side of the neck part of specimen may cause bending deformation. In the latter fatigue stage of $N_P > 10^3$, the residual stress is released by crack initiation, leading to the predominant shear deformation (Fig. 2.11).

## 2.4 Conclusion

The influence of fatigue damage on the elastic response of aluminum alloy has been investigated through acoustic wave velocity measurement and visualization of macroscopic deformation behavior using ESPI. In the earlier fatigue stage of pre-fatigue cycle $N_P < 10^3$, the change in the acoustic velocities suggested an increase of compressive residual stress along the tensile direction induced by the localized plastic deformation. On the other hand, the visualization of macroscopic deformation using ESPI demonstrated that strain heterogeneity in the macroscopic elastic regime was enhanced with increasing the pre-fatigue cycle, $N_P$. We infer that the residual stress induced by the fatigue cyclic load influenced ~~to~~ the macroscopic deformation behavior. These results indicate that the fatigue damage at the earlier fatigue stage due to the localized plastic deformation can be detected by the macroscopic deformation behavior using ESPI.

## References


1. Ogura, K., Miyoshi, Y., Kayama, M.: A study of X-ray analysis of fatigue fracture surface. Eng. Fract. Mech. **22**, 123 (1985)
2. Steuwer, A., Edwards, L., Pratihar, S., Ganguly, S., Peel, M., Fitzpatrick, M.E.: In situ analysis of cracks in structural materials using synchrotron X-ray tomography and diffraction. Nucl. Instrum. Methods Phys. Res. B. **246**, 246 (2006)
3. Steuwer, A., Rahman, M., Shterenlikht, A., Fitzpatrick, M.E., Edwards, L., Withers, P.J.: The evolution of crack-tip stresses during a fatigue overload event. Acta Mater. **58**, 4039 (2010)
4. Moorth, V., Jayakumar, T., Raj, B.: Influence of microstructure on acoustic emission behavior during stage 2 fatigue crack growth in solution annealed, thermally aged and weld specimens of AISI type 316 stainless steel. Mater. Sci. Eng. **A212**, 212 (1996)
5. Hasegawa, S., Sasaki, T., Yoahida, S., Hebert, S.L.: Analysis of fatigue of metals by electronic speckle pattern interferometry. Conf. Proc. Soc. Exp. Mech. **3**, 127 (2014)
6. Sasaki, T., Hasegawa, S., Yoahida, S.: Fatigue Damage Analysis of Aluminum Alloy by ESPI. Conf. Proc. Soc. Exp. Mech. **9**, 147 (2015)
7. Stratoudaki, T., Ellwood, R., Shrples, S., Clark, M., Somekh, M.G.: Measurement of material nonlinearity using surface acoustic wave parametric interaction and laser ultrasonics. J. Acoust. Soc. Am. **129**(4), 1721 (2011)
8. Rivière, J., Remillieux, M.C., Ohara, Y., Anderson, B.E., Haupert, S., Ulrich, T.J., Johnson, P.A.: Dynamic acousto-elasticity in a fatigue-cracked sample. J. Nondestruct. Eval. **33**, 216–225 (2014). doi:10.1007/s10921-014-0225-0
9. Su, Z., Zhou, C., Hong, M., Cheng, L., Wang, Q., Qing, X.: Acousto-ultrasonics-based fatigue damage characterization: linear versus nonlinear signal features. Mech. Syst. Signal Process. **42**, 25 (2014)
10. Eira, J.N., Vu, Q.A., Lott, M., Payá, J., Garnier, V., Payan, C.: Dynamic acousto-elastic test using continuous probe wave and transient vibration to investigate material nonlinearity. Ultrasonics. **69**, 29 (2016)
11. Yoshida, S., Sasaki, T., Craft, S., Usui, M., Haase, J., Becker, T., Park, I.-K.: Stress analysis on welded specimen with multiple methods. Conf. Proc. Soc. Exp. Mech. **3**, 143 (2015)
12. Toda, H., Fukuoka, H., Aoki, Y.: R-value acoustielastic analysis of residual stress in a seem plate. Jpn. J. Appl. Phys. **23**, 86 (1984)
13. Yoshida, S., Miura, F., Sasaki, T., Rouhi, S.: Optical analysis of residual stress with minimum invasion. In: Conference and Exposition on Experimental and Applied Mechanics, Indianapolis, USA, #141 (2017)


**Tomohiro Sasaki** Associate professor, The topics includes fatigue analysis, measurement of welding induced residual stress metals, using optical and acoustical techniques.

# Chapter 3
# Early Strain Localization in Strong Work Hardening Aluminum Alloy (2198 T3): 3D Laminography and DVC Measurement

**Ante Buljac, Lukas Helfen, François Hild, and Thilo F. Morgeneyer**

**Abstract** The effect of strain hardening on localization in front of a notch is assessed by following the interactions between strain concentrations, damage, initial microstructure and grain orientations. A CT-like specimen made of strong work hardening 2198 T3 aluminum alloy is subjected to an *in situ* synchrotron laminography experiment. Kinematic fields are measured via digital volume correlation. The final results are bulk displacement and strain fields including their corresponding resolutions. The reported results refer to the portion of the specimen around 1 mm away from the notch root. With the selected spatial resolution, damage nucleation and growth is evaluated in strained bands until the very end of the loading process.

**Keywords** Digital Volume Correlation (DVC) • Flat-to-slant transition • High work hardening material • Laminography • Plastic flow

## 3.1 Introduction

One of the examples that summarize well the current challenges in ductile damage understanding and modeling is ductile tearing. It has been reported [1–6] that during mode I opening of Compact Tension (CT) specimens, the initial crack starts to propagate normal to the loading direction but then tilts and continues in a slant manner as a combination of modes I and III. The investigation of the origin for such behavior has its engineering relevance since it has been shown that mixed-mode I/III leads to reduced toughness when compared to pure mode I fracture [7].

The parallel use of laminography [8] and global Digital Volume Correlation (DVC) [9] enables *in situ* displacement fields to be measured at the microscale. Laminography as a non-destructive imaging technique allows flat specimens to be analyzed (Fig. 3.1a) and wider stress states to be achieved when compared with tomography. Thanks to DVC, bulk displacement fields can be measured and strain fields calculated [10]. Up to now, each new portion of the results obtained by laminography and DVC reveals interesting phenomena at micrometer resolutions. This is *per se* a strong motivation for further analyses on other aluminium alloys. However, to be consistent it was decided to proceed in a systematic manner by selecting a specific family of alloys using different heat treatments. In particular, aluminum alloys AA2139 and AA2198 have been studied [10–14]. Representing the latest generation of aeronautical alloys, the understanding of the underlying failure mechanisms is of great interest. Extensive analyses on CT-like specimen made of AA2198 T8 [11, 13] have shown that in the zone close to the notch root, *i.e.* ~800 μm from the notch [11] and even in its immediate vicinity [13], strained bands appeared early in

A. Buljac (✉)
Laboratoire de Mécanique et Technologie (LMT), ENS Paris-Saclay / CNRS / Université Paris-Saclay, 61 avenue du Président Wilson, 94235, Cachan Cedex, France

MINES ParisTech, PSL Research University, Centre des Matériaux, CNRS UMR 7633, BP 87, 91003, Evry, France
e-mail: buljac@lmt.ens-cachan.fr

L. Helfen
ANKA/Institute for Photon Science and Synchrotron Radiation, Karlsruhe Institute of Technology (KIT), 76131, Karlsruhe, Germany

European Synchrotron Radiation Facility (ESRF), 38043, Grenoble, France

F. Hild
Laboratoire de Mécanique et Technologie (LMT), ENS Paris-Saclay / CNRS / Université Paris-Saclay, 61 avenue du Président Wilson, 94235, Cachan Cedex, France

T.F. Morgeneyer
MINES ParisTech, PSL Research University, Centre des Matériaux, CNRS UMR 7633, BP 87, 91003, Evry, France

 
L. Lamberti et al. (eds.), *Advancement of Optical Methods in Experimental Mechanics, Volume 3*,
Conference Proceedings of the Society for Experimental Mechanics Series, DOI 10.1007/978-3-319-63028-1_3

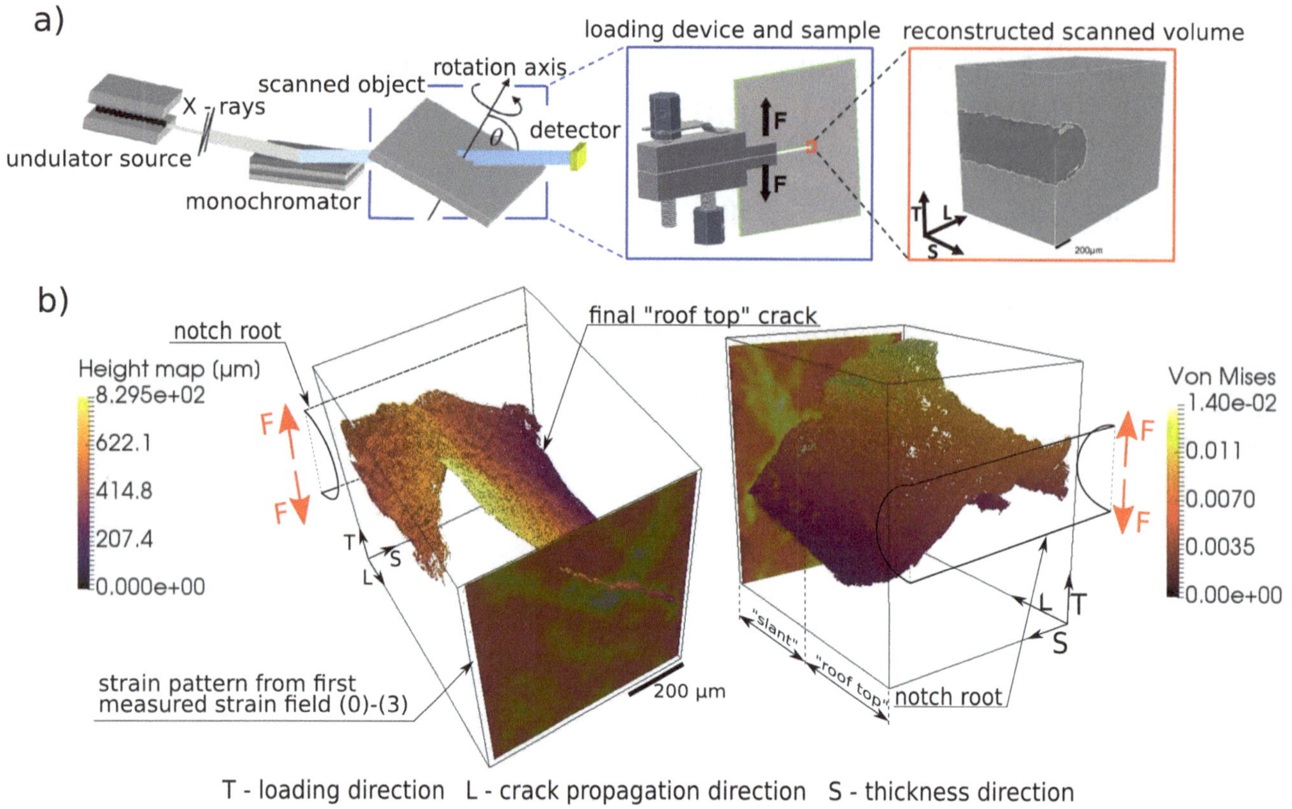


**Fig. 3.1** (**a**) Schematic representation of the laminography technique with the loading device and CT-like sample; (**b**) early strain field from slanted region shown together with the position of the final failure illustrated as height map

the loading history. Plane strain conditions in the crack propagation direction predicted in numerical simulations [15] have been confirmed by kinematic measurements [13] while through thickness equivalent strain distributions varied significantly. Slant strained bands appeared well before damage, *i.e.* plastic flow is very heterogeneous before any significant signs of void nucleation and growth is measured at micrometer resolutions. The origin and observed behavior (*i.e.* activation and deactivation) of strained bands at the microscale still remains unclear.

The T8 heat treatment induces relatively high yield stress followed by low work hardening. Low work hardening materials are believed to be prone to flow localization [16]. Following the same logic, strong work hardening materials should be more resistant to the flow localization. However, this was not the case for another aluminum alloy (*i.e.* AA2139 T3). Numerous strain bands were observed. Their intermittent activity and pattern were two key findings [14]. Further, damage set in very late during loading. Since these two alloys were different, the question remains open for AA2139 with T8 treatment and AA2198 under T3 condition. The latter is studied herein.

## 3.2 Kinematic Fields and Damage Micromechanisms

In this work, results obtained for a CT-like specimen made of AA2198 T3 are presented. The T3 heat-treatment results in a microstructure responsible for a strong work hardening behavior, which is the key feature for conducting such analyses. Strong work hardening and high strains to failure indicate low level damage development and subsequent late softening phases. Logically, this should be promoted by homogeneous strain flow with less strain heterogeneities. Hence, the strain fields in this high work hardening material are analyzed in the work. Moreover, since similar analyses exist for the T8 heat-treatment [11], the influence of the latter is also studied.

The strain fields and damage micromechanisms in the slant region ∼1 mm from the notch root as well as in the region closer to the notch root are presented in the work. Despite strong work hardening, the slant strained bands occur early on in the loading history before any signs of damage onset. Interestingly, void growth inside and outside the bands does not differ significantly, yet the final crack path follows the band position. This does not happen without any damage changes since damage growth is observed at micrometer resolution but only very late in the loading history. Figure 3.1b illustrates early strain fields plotted on the section normal to the crack propagation direction together with the final crack position.

Qualitative observations and damage quantifications lead to the identification of two different damage micromechanisms, namely, classical high stress triaxiality induced and shear type damage growth. The latter plays a key role in the slant zone of the CT-like specimen, while the flat zone is characterized by the superposition of both micromechanisms. However, even in the flat zone the shear-type micromechanism prevails, which causes the zigzagged final crack pattern close to the notch root.

**Acknowledgments** The financial support of the Fédération Francilienne de Mécanique and Agence Nationale de la Recherche (ANR-14-CE07-0034-02 grant for COMINSIDE project) is gratefully acknowledged. *Constellium C-Tech* supplied the material. We acknowledge the European Synchrotron Radiation Facility for provision of beamtime at beamline ID19 (experiments MI-1149 and MA-1631).

## References

1. Mahgoub, E., Deng, X., Sutton, M.A.: Eng. Fract. Mech. **70**, 2527–2542 (2003)
2. Bron, F., Besson, J., Pineau, A.: Mat. Sci. Eng. **A380**(1–2), 356–364 (2004)
3. Asserin-Lebert, A., Besson, J., Gourgues, A.F.: Mat. Sci. Eng. **A395**(1–2), 186–194 (2005)
4. Lan, W., Deng, X., Sutton, M.A., Chen, C.S.: Int. J. Fract. **141**, 469–496 (2006)
5. Morgeneyer, T.F., Helfen, L., Sinclair, I., Proudhon, H., Xu, F., Baumbach, T.: Scripta Mat. **65**, 1010–1013 (2011)
6. Besson, J., McCowan, C.N., Drexler, E.S.: Eng. Fract. Mech. **104**, 80–95 (2013)
7. Kamat, S.V., Hirth, J.P., Eng, J.: Mater. Technol. **117**(4), 391–394 (1995)
8. Helfen, L., Baumbach, T., Mikulík, P., Kiel, D., Pernot, P., Cloetens, P., Baruchel, J.: Appl. Phys. Lett. **86**(7), 071915 (2005)
9. Roux, S., Hild, F., Viot, P., Bernard, D.: Comp. Part A. **39**(8), 1253–1265 (2008)
10. Morgeneyer, T.F., Helfen, L., Mubarak, H., Hild, F.: Exp. Mech. **53**(4), 543–556 (2013)
11. Morgeneyer, T., Taillandier-Thomas, T., Helfen, L., Baumbach, T., Sinclair, I., Roux, S., Hild, F.: Acta Mat. **69**, 78–91 (2014)
12. Taillandier-Thomas, T., Roux, S., Morgeneyer, T.F., Hild, F.: Nucl. Inst. Meth. Phys. Res. B. **324**, 70–79 (2014)
13. Buljac, A., Taillandier-Thomas, T., Morgeneyer, T.F., Helfen, L., Roux, S., Hild, F.: Int. J. Fract. **200**(1), 49–62 (2016)
14. Morgeneyer, T.F., Taillandier-Thomas, T., Buljac, A., Helfen, L., Hild, F., Mech, J.: Phys. Solids. **96**, 550–571 (2016)
15. Bron, F., Besson, J.: Eng. Fract. Mech. **73**(11), 1531–1552 (2006)
16. Rice, J.R.: Theoretical and Applied Mechanics, pp. 207–220 (1976)

**Ante Buljac** was born 17th November 1989 in Sinj, Croatia. He finished gymnasium in Sinj and Faculty of Mechanical Engineering and Naval Architecure in Zagreb, Croatia. At the time being, he is 3rd year phd student and works for Laboratoire de Mécanique et Technologie (LMT) Cachan and Centre des Matériaux (CdM) Evry under COMINSIDE project founded by french Agence Nationale.

# Chapter 4
# On the In-Plane Displacement Measurement by 3D Digital Image Correlation Method

**Chi-Hung Hwang, Shou Hsueh Wang, and Wei-Chung Wang**

**Abstract** In this study, a three-step object was moved in-plan to different positions and determined the displacement by 3D DIC method. To monitor the out-of-plane displacement of the object during movement, a laser triangulation method was performed together with a typical 3D DIC. The laser triangulation setup consists of a laser source with a cylindrical lens mounted on the center of the camera-pair to generate a line pattern and then the discontinuous line patterns were then recorded by left- and right- cameras for step-height calculation. The measurement results reveal that there is displacement deviation introduced by viewing angle between camera and the object as the object is positioned at different locations. The experiment also shows that the deviation of 3D DIC determined in-plane displacement is small but the determined displacement is higher than the nominal displacement determined by a precision stage; in addition, out-of-plane displacements have been reported and cannot be ignored. In the final, a single camera model based on geometrical parameters of 3D DIC is proposed to correct the 3D DIC determined displacements; according to current results, the in-plane displacement of the 3-steps object can be tremendously reduced from 9.7% to 1.7%.

**Keywords** 3D DIC • In-plane displacement • Laser triangulation method • Different depths • Step high

## 4.1 Introduction

Digital image correlation (DIC) method becomes essential optical measurement method because DIC method is one of noncontact measurement methods which can be implemented for industrial and other on-site applications. The principle of DIC method is based on tracking artificial or natural surface random pattern among series captured images, therefore, two images from different status are needed for evaluating the change from one frame to the others, which means the associated displacement, strain and stress can be evaluated from two images captured or extracted from the recorded video stream. Moreover, the "image" is not limited to the traditional picture captured by the optical lens; for DIC method, the so called "image" has been successfully extended to two dimensional/ three dimensional spatial digital data which are obtained by SEM, TEM, AFM and XCT for micro-scale, nano-scale and volume object respectively. DIC methods can be classified into 2D DIC and 3D DIC method according to the optical system used for capturing optical images for analysis.

There are possible errors could be introduced by 2D DIC method unless surfaces of objects under testing are flat, the camera is perpendicular aligned with respect to the specimen and in-plane displacement [1–3]. In general, 3D DIC which is based on the stereo-photogrammetry has been considered as the solution to eliminate errors introduced by out-of-plane displacement [1, 3–5] because the visual triangulation introduced by stereo-images can provide extra data to reconstruct the depth information of the test object. Different from previous studies, in this paper, a three-step aluminum block is manufactured as a test object which was moved in-plane with well aligned precious linear stage along x-axis to evaluate the 3D DIC measurement accuracy of in-plane displacement through depth-direction. By using the images obtained by the camera-pair setup, distance from camera-pair to the aluminum block and the DIC software, the measurement results showed that the in-plane displacement between the 3D DIC determined one is extremely large than the nominal in-plane displacement, moreover, out-of-plane displacement was also presented. Be aware of this new challenge, the measurement differences were discussed and then a model was proposed to correct the 3D DIC determined in-plane displacement.

C.-H. Hwang (✉)
ITRC, NARL, Hsinchu, Taiwan, 30073, Republic of China
e-mail: chhwang@itrc.narl.org.tw

S.H. Wang • W.-C. Wang
Department of Power Mechanical Engineering, National Tsing Hua University, Hsinchu, Taiwan, 30013, Republic of China



L. Lamberti et al. (eds.), *Advancement of Optical Methods in Experimental Mechanics, Volume 3*,
Conference Proceedings of the Society for Experimental Mechanics Series, DOI 10.1007/978-3-319-63028-1_4

## 4.2 Experimental Setup

In this study, a typical 3D DIC configure composed with a laser triangulation setup were implemented for in-plane displacement measurement, as shown in Fig. 4.1. The laser triangulation setup consisted of a laser source and a cylindrical lens which were integrated as an unit and then mounted right at the center in between left- and right- cameras; based on the experimental setup, two cameras can be used to take image-pairs for determining in-plane displacement by using 3D DIC method; at the same time, the left- and right- cameras can also capture a bright curve introduced by the projected laser light-sheet intersects with the specimen. The heading angles of the left- and right- cameras are rotated to be about ±10.5° with respect to the specimen. The baseline distance of the two cameras is set to be about 188.6 mm.

A three-step aluminum block was used as the test specimen, the change of height indicates the geometrical change of the experimental setup which can be used to monitor the change of the geometrical setup. The sizes of the three-step specimen are 10 × 10 mm$^2$, 30 × 30 mm$^2$ and 60 mm × 70 mm$^2$ with respect to the front surface, the middle surface and the rear surface, and the offsets of front-to-middle surface and middle-to-rear surface are 10.18 mm and 20.25 mm respectively.

For 3D DIC in-plan study, the three-step-height specimen was moved by a manual linear-stage along x-axis away 3, 6, 9, 12, 15, 18, 21 and 24 mm with respect to its origin. As shown in Fig. 4.2, the displacements were extracted from three

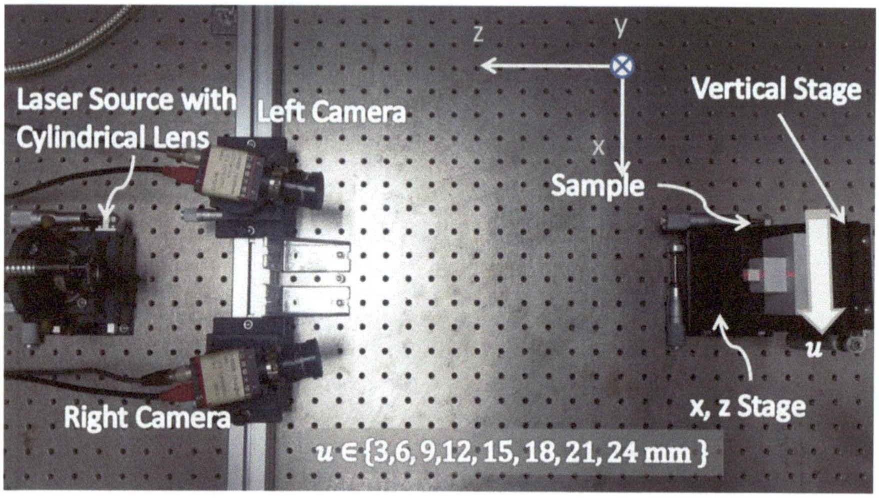


**Fig. 4.1** Experimental setup consists of the 3D DIC and the laser triangulation method

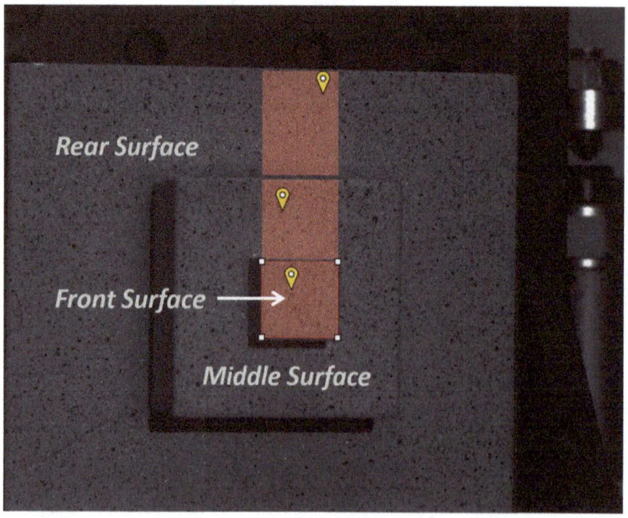


**Fig. 4.2** A three-step aluminum block is moved in plan along x-direction and the displacement is evaluated from remarked regions of the block by 3D DIC method

**Table 4.1** Calibrated parameters of camera-pair

| Parameters | $C_x$ | $C_y$ | $f_x$ | $f_y$ | *Skew* | $\kappa_1$ | $\kappa_2$ | $\kappa_3$ |
|---|---|---|---|---|---|---|---|---|
| Left camera | 783.22 | 525.51 | 8709.47 | 8707.76 | −1.4528 | −0.2015 | 0 | 0 |
| Right camera | 819.33 | 512.09 | 8718.95 | 8717.76 | −1.7855 | −0.2026 | 0 | 0 |

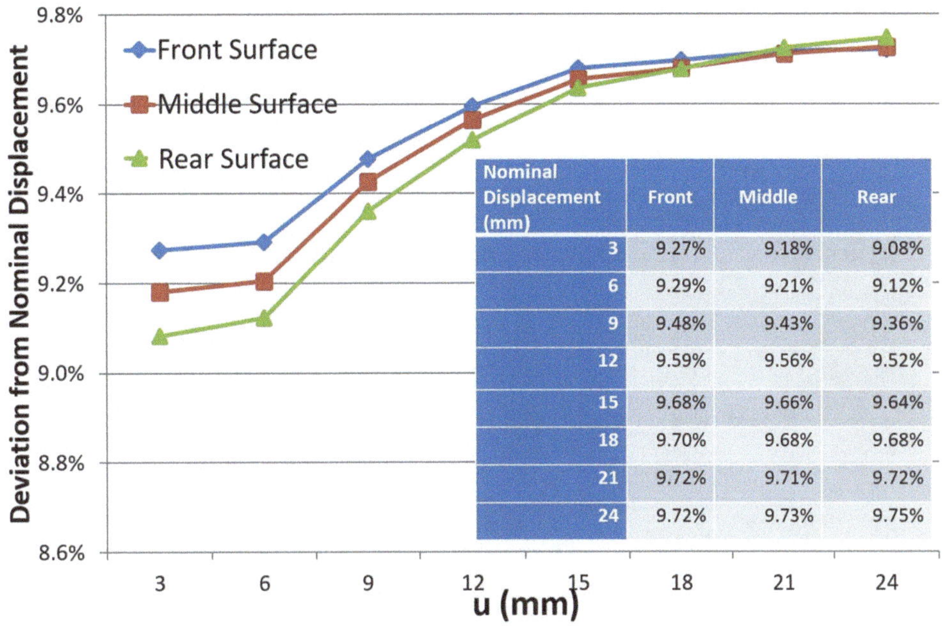


| Nominal Displacement (mm) | Front | Middle | Rear |
|---|---|---|---|
| 3 | 9.27% | 9.18% | 9.08% |
| 6 | 9.29% | 9.21% | 9.12% |
| 9 | 9.48% | 9.43% | 9.36% |
| 12 | 9.59% | 9.56% | 9.52% |
| 15 | 9.68% | 9.66% | 9.64% |
| 18 | 9.70% | 9.68% | 9.68% |
| 21 | 9.72% | 9.71% | 9.72% |
| 24 | 9.72% | 9.73% | 9.75% |

**Fig. 4.3** In-plan displacement derivations with respect to nominal displacement-raw data

surfaces, named as front, middle and rare surface, and averaged over the each area of each surface. The displacements were determined by using a commercial DIC system VIC 3D, the camera-pair has calibrated by proposed calibration procedures and the associated parameters were shown in Table 4.1.

## 4.3 Results and Discussions

In this study, the in-plane displacement was first determined by using 3D DIC with 27 × 27 pixels subset. Different from previous studies, a three-step specimen is used instead of a plate. Because of the surfaces are located at different depths, therefore, 3D DIC determined in-plane displacements can be obtained across the depths at the same time. Considering rotation along y-axis might be introduced while moving the test specimen along x-axis, a laser triangulation method was implemented in this study and the two available cameras captured the images from left- and right- views to monitor the movement. The detail measurement results are presented and discussed by the following sections.

### *4.3.1 In-Plane Displacements at Different Depths*

The determined in-plane displacements at different surfaces were illustrated in Fig. 4.3, the results showed that the deviations between the determined in-plan displacements and nominal displacements are always lager than 9%, and the deviations are increasing as the in-plane displacement enlarged. The illustrated Fig. 4.3 also revealed very interesting informations; the deviations from nominal in-plane displacement in magnitude are ordered as $\Delta u|_{\text{front surface}} > \Delta u|_{\text{middle surface}} > \Delta u|_{\text{rear surface}}$ when the object was moved from 0 to 21 mm; in fact, the deviation difference among three surfaces at a given nominal in-plane displacement also decreases when the object is moved from 0 to 21 mm and the deviation is almost the same for all three surfaces as the nominal in-plane displacement is 21 mm in this study.

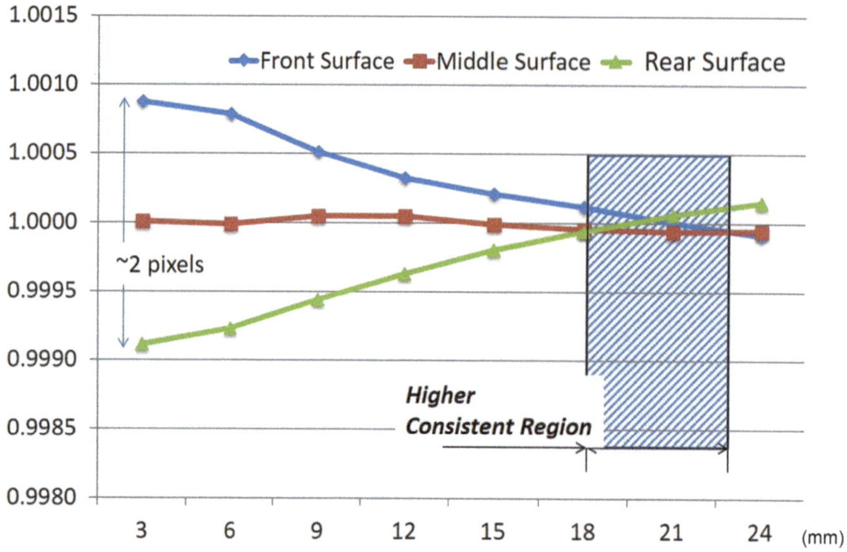


**Fig. 4.4** The ratios of 3D DIC determined in-plan displacement *u* w.r.t averaged in-plane displacement

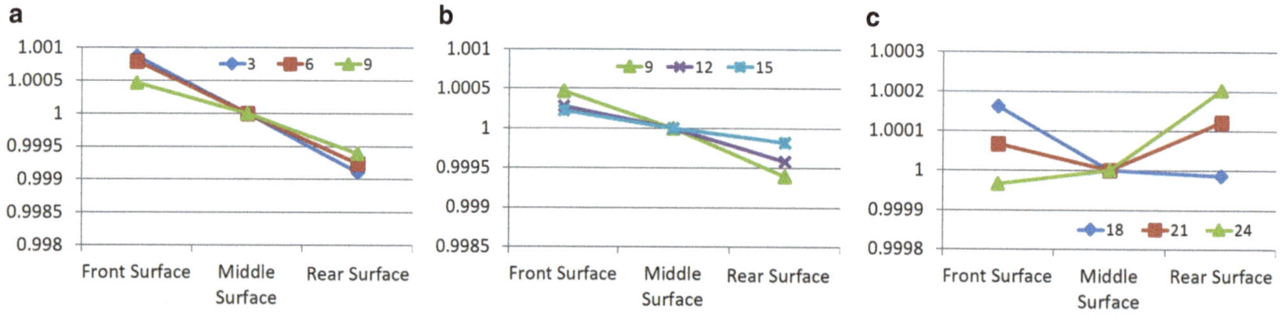


**Fig. 4.5** Normalizing in-plan displacement *u* of different surface w.r.t 3D DIC determined displacement of middle surface. (**a**) Moving towards +x- axis to 3, 6 and 9 mm. (**b**) Moving towards +x- axis to 9, 12 and 15 mm. (**c**) Moving towards +x- axis to 18, 21 and 24 mm

Figure 4.4 is the illustration of the determined in-plane displacements of all three surfaces are compared with the averaged values. The results show that the maximum deviation is about 2 pixels which happened to be moving the aluminum block 3 mm away from its origin and then decreased as the displacement increased until the in-plane displacement is 21 mm. It should be emphasize that the deviation-differences can be identified however the difference of the determined in-place displacement in magnitude among three surface are small. Moreover, inspecting Fig. 4.4 carefully, it could be fund that the averaged displacement is closed to the middle surfaces because the middle surface-to-averaged in-plane displacement is 1 with small fluctuation which means the 3D DIC evaluated in-plane displacement can be approximated by the data of middle surface. In addition, Fig. 4.4 also implied that the determined in-plane displacement might be a function of depth. Pleas also be award that the determined in-plane displacement is over 9% higher than the nominal one, therefore, it is important to know that the determined in-plane displacement near the camera-pair is higher/ lower than the determined displacement related far from the camera-pair, by this way the measurement can be improved by increasing/ decreasing the distance between the camera-pair and object or additional in-plane displacement calibration should be performed to ensure the distance between the camera-pair and the testing object is proper.

Since the middle surface can be used as a reference of measurement, the measured in-plane displacement of rear and front surfaces are normalized with respect to the middle surface one and the results are shown in Fig. 4.5. By comparing the plotted curves, interesting to know, in this study, that the ratio can be described as a linear function for small in-plane displacement, in this study, from 3 to 15 mm. The slope of the estimated linear functions decreased in magnitude as the in-plane displacement was increased, and then when the determined in-plane displacement is large then 15 mm, then the determined in-plane displacement from front- to rear- surfaces cannot be considered as linear; eventually, the determined in-plane displacement of rear- surface become higher- than the front surface.

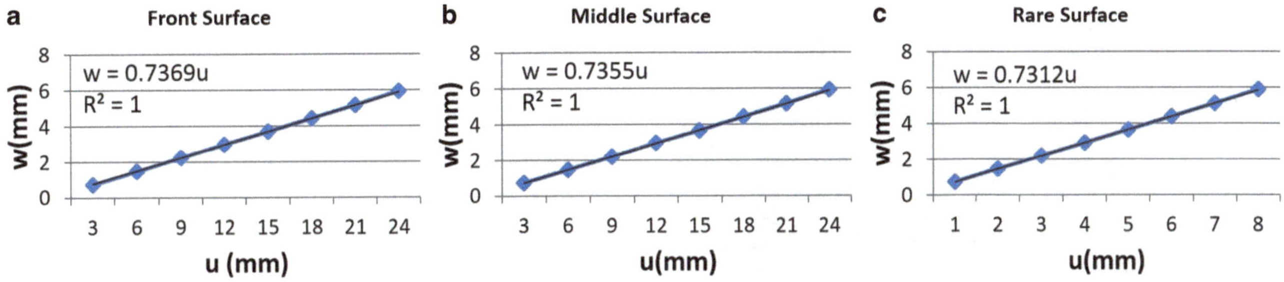


**Fig. 4.6** Associated out-of-plane displacement by 3D DIC as the object is moved in-plane. (**a**) Front surface. (**b**) Middle surface. (**c**) rare surface

**Table 4.2** $(\Delta\overline{w})/(\Delta\overline{u})$ verus nominal displacement $u$

| Nominal Displ. $u$ | Averaged Displacement | | | |
|---|---|---|---|---|
| | $\overline{\Delta w}$ | $\overline{\Delta u}$ | $\overline{\Delta w}/\overline{\Delta u}$ | $\theta = \tan^{-1}\frac{\overline{\Delta w}}{\overline{\Delta u}}$ |
| 3 | – | – | – | – |
| 6 | 0.7301 | 3.2770 | 0.2228 | 12.5604 |
| 9 | 1.4623 | 6.5725 | 0.2225 | 12.5430 |
| 12 | 2.1975 | 9.8718 | 0.2226 | 12.5499 |
| 15 | 2.9278 | 13.1732 | 0.2223 | 12.5308 |
| 18 | 3.6746 | 16.4677 | 0.2231 | 12.5790 |
| 21 | 4.4183 | 19.7653 | 0.2235 | 12.6006 |
| 24 | 5.1615 | 23.0602 | 0.2238 | 12.6164 |
| Averaged angle $\theta$ | | | | 12.5686 |

### 4.3.2 *Out-of-Plane Displacement*

Although the testing object was moved along x-axis, out-of-plan displacement was obtained from 3D DIC analyzed results. In this study, the out-of-plane displacement was increased as the in-plane displacement increased. As shown in Fig. 4.6, the out-plane displacement and in-plane displacement can be well fitted by a linear function. The slope of the fit curves are 0.7369, 0.7355 and 0.7312 with respect to front-, middle- and rear- surfaces and the corresponding $R^2$ can be approximated to be 1. In addition, the slope of u-w plots are fitted with respect to the step-height difference by using front-surface as a reference, again, the step-high difference (Δh) versus u-w slope (s) can be described as $s = -60 \times 10^{-6}\,\Delta h + 0.2457$ with corresponding $R^2 = 0.9822$.

Table 4.2 presents a very interesting result regarding the averaged in-plane displacement $(\Delta\overline{u})$ and the averaged out-of-plane displacement $(\Delta\overline{w})$ with 3 mm nominal averaged in-plane and out-of-plane displacement as reference. The results show that the calculated $(\Delta\overline{w})/(\Delta\overline{u})$ values are always about 0.2223 which means the increment of $w$ is highly dependent with u. Meanwhile the $(\Delta\overline{w})/(\Delta\overline{u})$ ratio can be converted to an angle defined by u and w. The angle is about 12.5686° and this angle is close to the 10.5° heading angle of camera. In fact, the estimated angle defined by displacement u and w could be the implemented heading angle of the camera used for 3D DIC, however, more evidences based on experimental results before conclusion.

### 4.3.3 *Laser Triangulation Measurement*

As indicated in 3.2, the out-of-plane displacement can be evaluated by 3D DIC while the object was moved in-planes along x-axis. Although the movement is performed by using a precision stage, however, monitoring the possible out-of-plane displacement caused by geometrical misalignment of camera-pair and testing object is essential. In this study, a laser and cylindrical lens integrated light source was moved together with the aluminum block, the step height of the aluminum block was then evaluated from left- and right- cameras; by this way, if there is unparallel available between the precious stage and base-line of camera-pair, then the unparallel can be detected because of the change of height between steps of the specimen. The measurement results were shown in Fig. 4.7, the height difference between front- and middle- surfaces obtained from left- and right- cameras are almost identical when the displacement is less than 10 mm; but the difference of evaluated

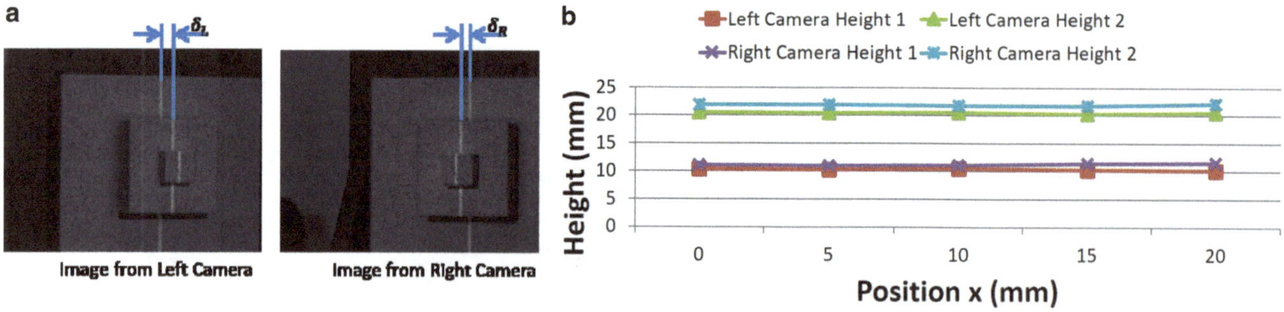


**Fig. 4.7** The depths of an aluminum block evaluated by using laser triangulation methods at different x-positions. (**a**) Typical images obtained from Left- and Right cameras used for depth evaluation by laser triangulation method. (**b**) The depths observed from left- and right- cameras as the object is moved along +x-direction to different positions

height were enlarged as the in-plane displacement increased. Meanwhile, the step-height between front- and middle- surface determined from left- camera is slightly decreased as the in-plane displacement increased, but the height determined by the right-camera is converse. As for the determined height between middle- and rear- surfaces, the determined heights from both left- and right- cameras at different locations along x-axis are departure from each other, however the offsets are almost the same in this study.

In this study the out-of-plane runoff caused by moving the object along the linear stage should be ignored. Considering the determined out-of-plane displacement is monotonously increased as the in-plane displacement increased which means there is a small angle introduced by the precision stage and the structure used to support the camera-pair. However, from laser triangulation measurement results, the height of middle- and rear-surface remain the same value as the object move from 0 to 20 mm away from its origin; the heights of front- and middle- surface were different at different locations along x-axis and the maximum difference is about 1.36 mm which is small than the corresponding out-of-plane displacement determined by 3D DIC.

### *4.3.4 Correcting the In-Plane Displacement*

According to the measurement results presented previously, the evaluated in-plane displacement is 9% above the nominal in-plan displacement and the out-of-plane displacement is available with about 1/4 in-plane displacement in magnitude. In addition, according to laser triangulation measurement, the aluminum block was moved by a precision stage should be moved toward x-axis parallel to the camera-pair baseline; that means there is no significant out-plane displacement will be introduced as the measurement is executed. Since the angle defined by in-plane and out-of-plane displacement is almost a constant all over the displacement range, therefore, the deviation of in-plane displacement and the associated unforeseen out-of-plane displacement are considered to be introduced by the cameras layout. To correct the deviation, a single pinhole-camera model with half-baseline distance and the heading angle with respect to the object is 12.5686° which is introduced by $\Delta\overline{u}-\Delta\overline{w}$ to simplify the camera-pair implemented for 3D DIC imaging system. Then, in this study, the in-plane displacement can be corrected by the following equation,

$$u \sim \frac{\overline{b}+\overline{u}}{\cos(\phi+\varphi)}\cos(\theta+\varphi+\phi)-b \tag{4.1}$$

Where $\theta$ is the angle defined by $\Delta\overline{u}-\Delta\overline{w}$, $\varphi$ is angle spand by half baseline of the camera-pair with angle vertex at pinhole, $\phi$ is the angle spand by in-plane displacement along x-axis with its vertex locates at pinhole, $b$ is the half baseline of camera-pair, $\overline{b}$ is defined to be distance of baseline $b$ projecting with angle $\varphi$ along the direction of $\Delta\overline{u}$. Assuming $b$ and distance between pinhole and object are known, then the angle $\varphi$ and $\overline{b}$ can be determined. As shown in Fig. 4.8, the blue line is the in-plane displacement deviation with respect to the nominal one and the red line is the in-plane displacement deviation which is corrected by Eq. (4.1). The results show that the maximum difference in magnitude between the nominal displacement and evaluated in-plane displacement before and after correcting both occur when the block aluminum is moved

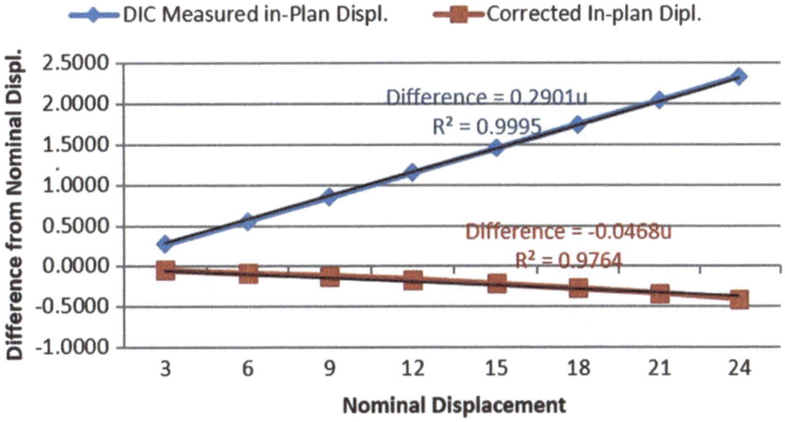


**Fig. 4.8** Different from nominal in-plane displacement before/after proposed geometrical correction model

24 mm away from its original location. Before correcting, the evaluated in-plane displacement is monotonously increased as the nominal in-plane displacement increased and the maximum difference is up to 2.3 mm larger than the nominal one which leads to 9.7%; controversy, the modified in-plane displacement is a little below the nominal in-plane displacement and the difference is monotonously decreased as the nominal in-plane displacement increased and the maximum difference is about 1.7% which is equivalent to 0.41 mm.

## 4.4 Conclusions

In this study, an in-plane displacement of a three-step aluminum block is determined by using 3D DIC method. The measured in-plane displacement in magnitude is much higher than the nominal in-plane displacement which is performed by moving the aluminum block with a precious linear stage. The difference between the 3D DIC determined in-plane displacement and nominal in-plane displacement was reported to be monotonously increased as the nominal in-plane displacement increased and the maximum difference is about 9.7% at nominal in-plane displacement is 24 mm. In addition, thanks to the aluminum block consists of three surfaces at different depths, the change of in-plane displacement at different depths can be determined at the same time. The results in this study present that the displacement determined by front- surface which is close to the camera-pair is highest for small displacement and the corresponding in-plane displacement of rear-surface, far away from the camera, is smallest. The results indicate the 3D DIC determined in-plane displacement is sensitive to the distance between the object and camera-pair. Meanwhile, from the 3D DIC evaluated displacement field, out-of-plane displacement was also reported. Again, the out-of-plane is monotonously increased as the nominal in-plane displacement increase. The ratios of the 3D DIC determined in-plane displacements and out-of-plane displacements at different nominal in-plane displacements are almost constant, in this study, which implied that the precious linear stage might be unparallel with respect to the frame used to support camera-pair for capturing the stereo-images or a no revealed characteristic of 3D DIC method. To identify the possible reason, a laser triangulation method was implemented with left- and right- camera was used to imaging the laser patterns on the aluminum block. The results obtained by using laser triangulation method indicate that the aluminum block is moved along precious linear stage without introduced out-of-plane displacement with respect to the geometrical setup of the camera-pair, therefore, according to current evidence, the out-of-plane displacement could be a characteristic 3D DIC method and the way to correct the 3D DIC determined in-plane displacement is essential. Based on a single camera model, a method to correct 3D DIC determined in-plane displacement was presented. The maximum difference between the determined in-plane displacement and the nominal in-plane displacement is improved from 9.7% to 1.7% in ratio and the corresponding displacements are 2.30 and 0.41 mm.

**Acknowledgements** The study is performed thanks to the financial support provided by the Ministry of Science and Technology of Taiwan, R.O.C. (Grant No. NSC-102-2221-E-492-014 and MOST-103-2221-E-492-017) is greatly appreciated.

## References

1. Sutton, M.A., Yan, J.H., Tiwari, V., Schreier, H.W., Orteu, J.J.: The effect of out-of-plane motion on 2D and 3D digital image correlation measurements. Opt. Lasers Eng. **46**(10), 746–757 (2008)
2. Haddadi, H., Belhabib, S.: Use of rigid-body motion for the investigation and estimation of the measurement errors related to digital image correlation technique. Opt. Lasers Eng. **46**(2), 185–196 (2008)
3. Lava, P., Coppieters, S., Wang, Y., Houtte, P., Van Debruyne, D.: Error estimation in measuring strain fields with DIC on planar sheet metal specimens with a non-perpendicular camera alignment. Opt. Lasers Eng. **49**(1), 57–65 (2011)
4. Gao, Y., Cheng, T., Su, Y., Xu, X., Zhang, Y., Zhang, Q.: High-efficiency and high-accuracy digital image correlation for three-dimensional measurement. Opt. Lasers Eng. **65**, 78–80 (2015)
5. Sutton, M.A., Orteu, J.-J., Schreier, H.: Image Correlation for Shape, Motion and Deformation Measurements: Basic Concepts, Theory and Applications. Springer, New York (2009)

**Dr. Hwang** works for Instrument Technology Research Center, NARLabs as a research fellow and now he also servers as Division Director of Applied Optics Division.

# Chapter 5
# Noise Reduction in Amplitude-Fluctuation Electronic Speckle-Pattern Interferometry

**Sanichiro Yoshida, David Didie, Jong-Sung Kim, and Ik-Keun Park**

**Abstract** Amplitude-Fluctuation Electronic Speckle Pattern Interferometry is used in a variety of vibration analyses. The technique utilizes the fact that when the vibration frequency of the object is significantly higher than the frame rate of the imaging device, the interference term can be approximated by the lowest (0th) order of the first kind Bessel function. Since the 0th order Bessel function takes the maximum value of unity when the vibration amplitude is null, the amplitude of a given vibration can be estimated from the reduction of the interference term relative to the case when the object is still. In reality, however, various environmental noises, such as temperature fluctuation of air in the interferometric paths and floor motion transferred through the optical table, cause low frequency fluctuations of the interference term, and thereby compromise the integrity of data. In this paper, we discuss typical environmental noise on standard optical interferometer settings, and propose to reduce the effect of the noise on the signal by introducing a carrier fringe system and analyzing the fringe pattern in the spatial frequency domain. The effectiveness of the proposed method is assessed for our recent Michelson interferometer experiment in which vibration of thin-film specimen is characterized.

**Keywords** Michelson interferometer • Amplitude-Fluctuation Electronic Speckle-Pattern Interferometry • Phase noise • Carrier Fringes • Thin-film adhesion

## 5.1 Introduction

Optical interferometry is widely used in various subfields of experimental mechanics. An interferometer is a devise to convert phase to intensity. The relative phase difference between the two interferometric arms is read out as a change in the intensity of the combined optical fields. Since there is no theoretical limitation in the phase resolution other than Heisenberg's uncertainty principle, the sensitivity in length measurement of an optical interferometer can be at the quantum physical level. This very advantage, however, makes an optical interferometer be vulnerable to environmental noise. It is important to deal with phase noise appropriately.

Another point of argument is the data acquisition time. Since light travels fast, the temporal behavior of the relative phase difference can be analyzed fast. This also means that a fast photo-detector is necessary to resolve the fast phenomenon. If the phase information is one-dimensional, i.e., simply associated with the arm length change, it is relatively easy to use a fast photo-diode such as a PIN-photo-diode. However, if the phase information is to be analyzed two-dimensionally, e.g., a full-field analysis of fast deformation of an object, often the data acquisition speed becomes an issue. A digital camera is most likely used to record the optical intensity that contains the phase information. Recent technological advances make high frame rate digital cameras commercially available. It is possible to obtain a digital camera with a frame rate of the order of $10^6$ frames per second or higher. However, the price of these fast devices is often beyond our budget, and in addition, there are still technical restrictions such as the limitation in the area of view in exchange for a high frame rate.

When an oscillatory phenomenon is to be analyzed with a digital camera whose frame rate is significantly lower than the oscillation frequency, Amplitude-fluctuation Electronic Speckle Pattern Interferometry (AF-ESPI) [1, 2] is useful. Basically, this technique estimates the oscillation amplitude of the test object from the interferometric fringe pattern formed in a two-dimensional, full-field image. Although the acquired image is time-averaged over the exposure time of the digital camera,

S. Yoshida (✉) • D. Didie • I.-K. Park
Department of Chemistry and Physics, Southeastern Louisiana University, SLU 10878, Hammond, LA, 70402, USA
e-mail: syoshida@selu.edu

J.-S. Kim
Department of Chemistry and Physics, Southeastern Louisiana University, SLU 10878, Hammond, LA, 70402, USA

Department of Mechanical and Automotive Engineering, Seoul National University of Science and Technology, Nowon-gu, Seoul, South Korea

 
L. Lamberti et al. (eds.), *Advancement of Optical Methods in Experimental Mechanics, Volume 3*,
Conference Proceedings of the Society for Experimental Mechanics Series, DOI 10.1007/978-3-319-63028-1_5

the nodes of the oscillation can be identified as dark fringes as the displacement on the node is always zero. By subtracting an image taken at a time step from the one taken another time step, we can form dark fringes at the nodes. If the frame rate is synchronized with the oscillation frequency, the bright fringe has a constant intensity and thereby we can analyze the intensity profile of the fringe image quantitatively.

Low frequency phase noise compromises the quality of the fringe pattern generated by AF-ESPI. Here the word low frequency is used to mean phase noise comparable to the digital camera's frame rate. If the phase noise changes the phase of the interferometric image between the two time steps used for the fringe formation, the dark fringe location is shifted. We need some measure to remove or compensate such phase noise.

We previously developed an interferometric technique to characterize the adhesion property of a thin-film to the substrate [3, 4]. The thin-film specimen was oscillated sinusoidally with an acoustic transducer and the resultant oscillation amplitude was read out interferometrically. With an algorithm analogous to AF-ESPI, the oscillation amplitude can be estimated from the contrast of the interferometric fringe pattern. In the course of this development, we noticed that low frequency noise due to environmental factors such as ambient air temperature fluctuation could considerably compromise the accuracy of the measurement. We also found that the analysis in the spatial frequency domain could significantly reduce the influence of low frequency phase noise.

The aim of this paper is to discuss low frequency phase noise and a way to reduce it. Data from the interferometric experiment for the above thin film adhesion characterization is used but the argument being made is applicable to other interferometric methods in general. The nature of low frequency phase noise is discussed in general. The reduction in low frequency phase noise by means of introduction of carrier fringes and frequency domain analysis is discussed.

## 5.2 Michelson Interferometer

### *5.2.1 One-Dimensional, Time Dependent Phase Analysis*

Figure 5.1 illustrates a typical Michelson-type interferometer. Laser light is split with a beam splitter into the reference arm and signal arm. The laser beam in the reference arm is reflected off a mirror and the laser beam in the signal arm is reflected off the surface of a specimen (in this study a thin film coated on a silicon substrate). An acoustic transducer attached to the specimen at the rear surface oscillates the specimen sinusoidally at a frequency set by the function generator. A fast photo-diode placed behind the beam splitter can be used to detect the signal arm length change due to the film surface oscillation as the corresponding relative optical phase difference.

The light intensity behind the beam splitter can be expressed as follows.

$$I(t) = 2I_0 + 2I_0 \cos\left[k\left(l_{so} - l_{ro}\right) + kd \sin \omega t\right] = 2I_0 + 2I_0 \cos\left[\delta_o + \delta \sin \omega t\right] \tag{5.1}$$

Here $I_0$ is the intensity of the reference and signal beams, $k$ is the wave number of the laser light in (rad/m), $l_{so}$ and $l_{ro}$ are the initial (physical) length of the signal and reference arms, $\delta_o = l_{so} - l_{ro}$ is the arm length difference, $\delta = kd$ and $d$ are the oscillation amplitude of the specimen surface in (rad) and (m), and $\omega$ is the oscillation (driving) angular frequency of the specimen. Here the reference and signal beams are the non-interfering light beams in the reference and signal arms,

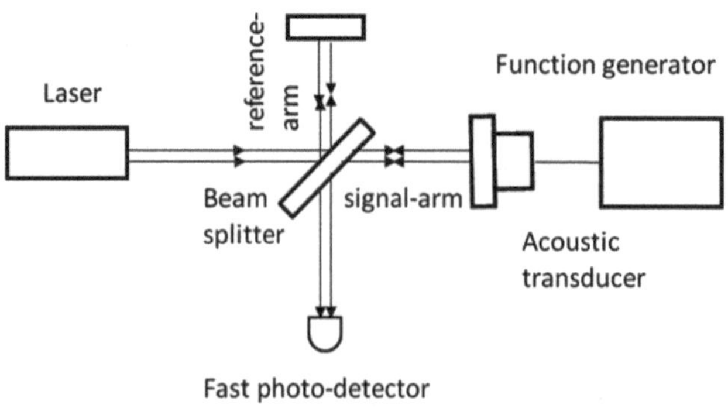


**Fig. 5.1** Michelson interferometer

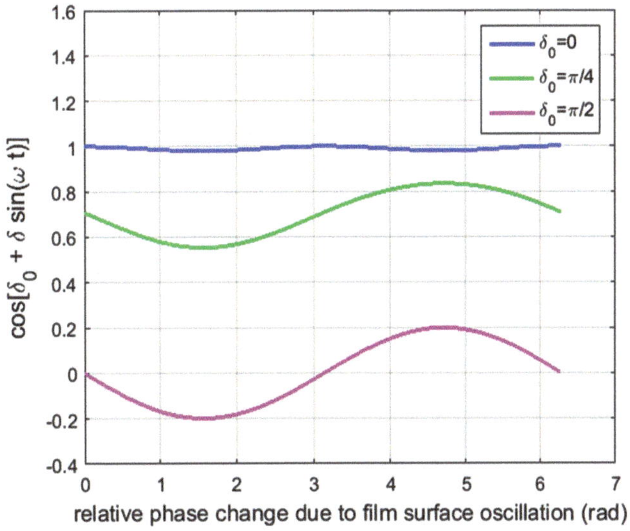


**Fig. 5.2** Interferometer sensitivity

respectively; their intensities are assumed to be equal to each other. (In reality they are not equal to each other but the gist of the argument here is not affected by inequality. The second term in the rightmost-hand side of Eq. (5.1) with $\cos(\delta_0 + \delta \sin \omega t)$ is called the interference term. This term is important as it contains the relative phase change information. The light intensity read by the photo-detector can be expressed as follows.

$$I_s(t) = I(t) - 2I_0 = 2I_0 \cos(\delta_o + \delta \sin \omega t) \tag{5.2}$$

Our purpose here is to extract $\delta = kd$ from $I_s(t)$, and thereby estimate the oscillation amplitude $\delta$. It is apparent that the sensitivity of the signal depends on the operation-point (initial) phase $\delta_o$. Figure 5.2 illustrates the signal sensitivity for three values of $\delta_o$. It is clear that when $\delta_o = \pi/2$, the sensitivity is the highest. The condition is that the arm length difference in the phase is a quarter wavelength, which means that the round trip phase difference between the two arms at the beam splitter is half wavelength. This condition is known as the destructive interference. Under this condition, when the amplitude $d = \delta/k = \delta/(2\pi\lambda)$ is null, the optical intensity is zero.

### 5.2.2 *Two-Dimensional Phase Analysis Based on Amplitude-Fluctuation Interferometry*

Figure 5.3 illustrates a typical setup for two-dimensional phase analysis. Out-of-plane displacement of a specimen is analyzed as a full-field image with the use of a digital camera. The dashed line indicates the case when the specimen is slightly tilted from the normal direction to the laser beam path from the laser so that a carrier fringe system is introduced. Call this case the carrier fringe configuration. We first discuss the other case where the specimen is normal to laser beam path (the solid line). Call this case the normal incident configuration.

In both cases, since the data acquisition time is significantly longer than the oscillation frequency, it is convenient to expand Eq. (5.1) into harmonics of the driving frequency and approximate the output signal with the time-independent components of the signal.

$$I(t) = 2I_0 \{1 + \cos(\delta_o + \delta \sin \omega t)\} = 2I_0 \{1 + \cos\delta_o \cos(\delta \sin \omega t) - \sin\delta_o \sin(\delta \sin \omega t)\} \tag{5.3}$$

We can rewrite Eq. (5.3) in terms of Bessel functions of the first kind as follows.

$$\begin{aligned} I(t) &= 2I_0 \cos\delta_o \{1 + J_0(\delta)\} \\ &+ 2I_0 \left[\cos\delta_o \left\{2J_2(\delta)\cos 2\omega t + 2J_4(\delta)\cos 4\omega t + \cdots t)\right\} - \sin\delta_o \{2J_1(\delta)\sin \omega t + 2J_3(\delta)\sin 3\omega t + \cdots\}\right] \end{aligned} \tag{5.4}$$

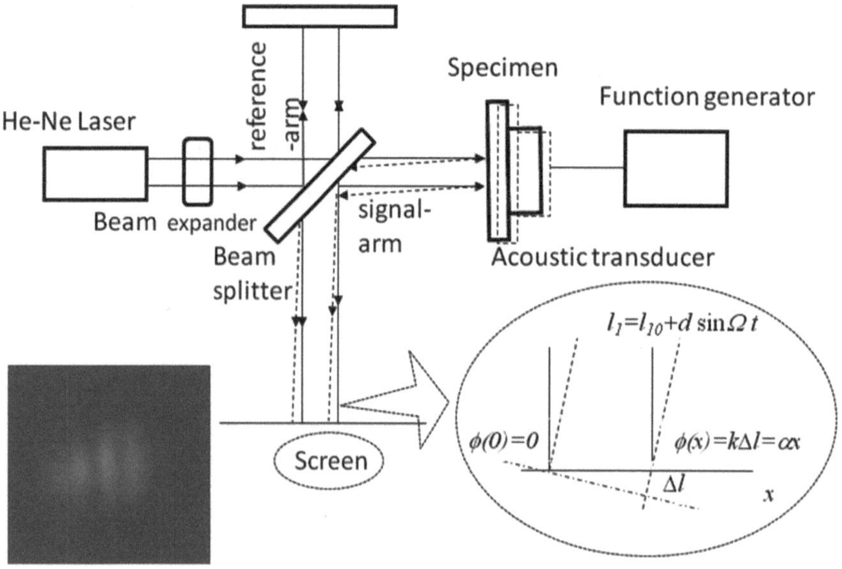


**Fig. 5.3** Carrier fringe configuration

When the above signal is taken with a digital camera whose exposure time is $\tau$, the output signal $S(\tau)$ is proportional to the time integration of $I(t)$. Here if the exposure time is significantly longer than the period of the oscillation $\omega$, i.e., $\tau >> 2\pi/\omega$, the oscillatory term on the right-hand side of Eq. (5.4) is much less than the first two non-oscillatory terms. By integrating the signal in the form of Eq. (5.4) for the exposure time, we obtain the following approximated expression of the signal.

$$S(\tau) = \int_0^\tau I(t)dt \cong 2I_0\tau\{1 + \cos\delta_o J_0(\delta)\} \tag{5.5}$$

Our purpose here is to extract $\delta = kd$ and thereby estimate the oscillation amplitude $\delta$. It is apparent that the sensitivity of the signal is proportional to the factor $\cos\delta_o$. Apparently, unlike the one-dimensional analysis with a fast photo-diode, the constructive interference condition $\delta_o = n\pi$ provides the highest sensitivity. This is naively understood as follows. In this case, the cosine function is integrated. The integration of a cosine function is essentially a sine function. Therefore, this time the phase providing the highest sensitivity is where the sensitivity is the lowest for the other case.

### 5.2.3 Phase Fluctuation due to Environmental Disturbance

In the normal incident configuration, the phase noise infiltrates into the $\cos\delta_o$ term as a function of time. If the frequency of the phase noise is low, there is no way to distinguish the change in the output signal $S(\tau)$ is due to $J_0(\delta)$ or deviation from the initial value of $\delta_o$ due to the phase noise unless the noise is characterized and the signal is high-pass filtered.

Before discussing the carrier fringe configuration, it is worth while estimating the magnitude of the low frequency phase noise under a realistic condition. When the interferometer is placed in air, temperature fluctuation plays a significant role. The optical phase change due to the temperature dependence of the refractive index of air can be expressed as follows.

$$d\varphi = 2\pi \frac{l}{\lambda}\frac{\partial n}{\partial T}dT \tag{5.6}$$

Here $\lambda$ is the wavelength, $l$ is the path length, $n$ is the refractive index of air and $dT$ is the temperature change. The temperature coefficient $\partial n/\partial T$ of air is $-0.87 \times 10^{-6}$ (1/°C) [5]. The arm length of the interferometer used in this experiment is 10 (cm). The wavelength of the laser used in this study is 632.8 nm. So, the phase change due to a temperature change of ±0.1 °C over the round trip in the interferometric arm is 20 (cm)/632.8 (nm) × 0.87 × 10–6 × 0.1 = 2.75% (of the period $2\pi$). Our measurement of air temperature in the interferometer arms indicate as much as 0.4 °C within a typical time of the interferometric experiment. The phase error due to this air temperature change is 2.75 × 4 = 11.0% of the period.

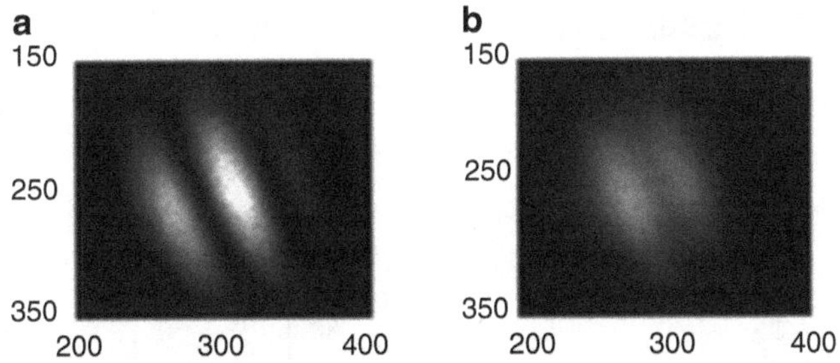


**Fig. 5.4** Sample fringe images (**a**) high contrast and (**b**) low contrast

### *5.2.4 Introduction of Carrier Fringe System*

Now we discuss the carrier fringe configuration (the case of the tilted specimen indicated with the dashed line in Fig. 5.3). In this case the factor $\cos\delta_o$ can be replaced by the spatial variation as follows.

$$S(\tau) \cong 2I_0\tau\left\{1+\cos\alpha x J_0(\delta)\right\} \tag{5.7}$$

Unlike the normal incident configuration, the operation-point phase varies through multiple periods in a given image; a dark fringe corresponds to $\cos\delta_o = \pi/2 + N\pi$ and the neighboring bright fringe to $\cos\delta_o = N\pi$ or $\cos\delta_o = (N+1)\pi$ ($N$: integer). The fluctuation of $\delta_o$, whatever its cause many be, changes the position of dark fringes or shifts the $x$ coordinate points that correspond to the condition $\cos\delta_o = \pi/2 + N\pi$.

In the space domain, the fringe shifts reduce the contrast. This indicates the possibility to correlate the fringe contrast reduction to the oscillation amplitude $\delta$. However, in the space domain, the correlation is not easy to quantify. Figure 5.4 illustrates typical fringe images of high and low contrasts. Instead, in the spatial frequency domain, it is easily quantified. We will discuss this in the next sections. It should be noted that low frequency phase noise causes the carrier fringes to shift in the space domain but as long as the angle of tilt made by the specimen remains the same the fringe spacing is intact. Therefore, in the frequency domain the peak frequency is unaffected by low frequency phase noise.

## 5.3 Experimental Results and Discussions

### *5.3.1 Normal Incident Configuration*

Figure 5.5 shows sample reference and signal beam intensity profiles of a two-dimensional phase analysis with the nominal incident configuration. It is seen that while the reference beam profile is Gaussian, the profile of the signal beam is deformed. The reason for the deformed profile is most likely that the specimen surface is not flat at the level of the laser wavelength and the reflectivity is not uniform.

Figure 5.6 shows the interference terms calculated by subtracting the sum of the reference and signal beam intensities (Fig. 5.5) from the signal intensity on a pixel-by-pixel basis. The reference beam and signal beam intensities were obtained by blocking one arm at a time. According to Eq. (5.5), the resultant quantity can be expressed as follows.

$$\frac{S(\tau)-2I_0\tau}{2I_0\tau} = \cos\delta_o J_0(\delta) \tag{5.8}$$

The two interference intensities were taken at two times steps 10 frames or approximately 300 ms apart. (Since the frame rate was 30 fps (frames per second), the separation in time for 10 frames is 10/30 = 0.333 s.) During this measurement, the acoustic driving frequency (hence $J_0(\delta)$) was unchanged. Therefore the difference between the two intensity profiles in Fig. 5.6 is due to the change in the phase $\delta_o$ in the $\cos\delta_o$ term.

The following observations can be made in regard to the two interference intensities. First they both are centered at $\cos\delta_o J_0(\delta) = 0.5$. This indicates that $\cos\delta_o$ happens to be 0.5, or $\delta_o = \cos^{-1}(0.5) = 1.05$(rad). Note that $J_0(0) = 1$ and those regions where the effect of the oscillation is negligible (i.e., the region far from the center of the beam) $\cos\delta_o J_0(\delta) = \cos\delta_o$. Second, the intensity profile at frame # 2 bulges more on the positive side whereas the one at frame # 12 bulges more on

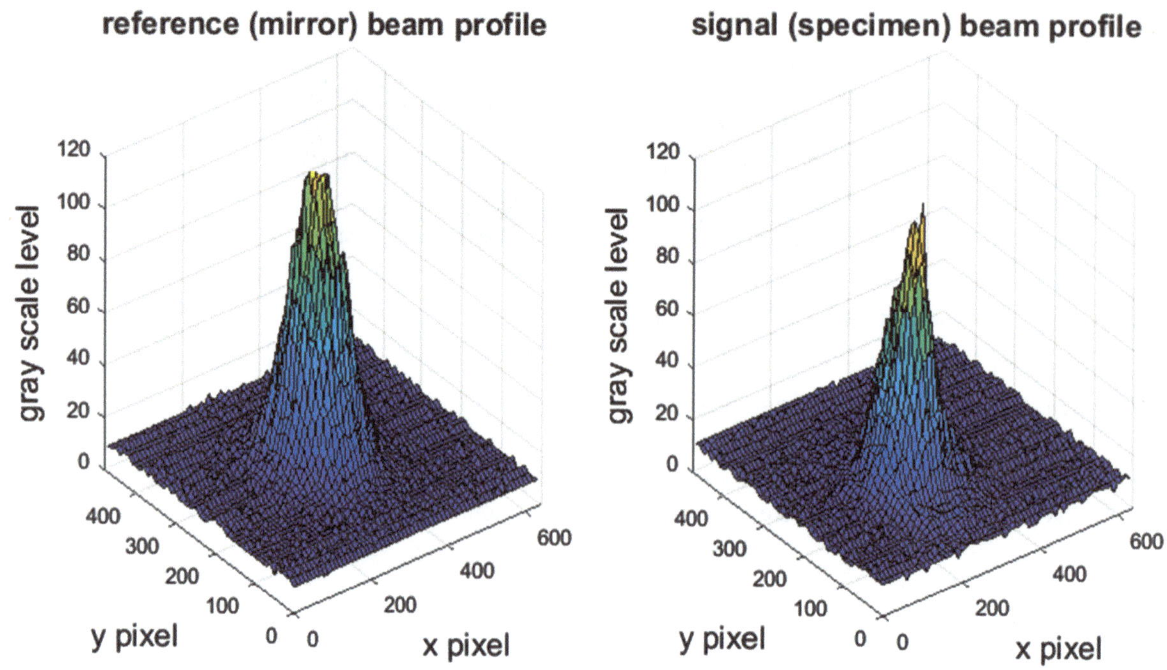


**Fig. 5.5** Intensity profile of the reference and signal beams

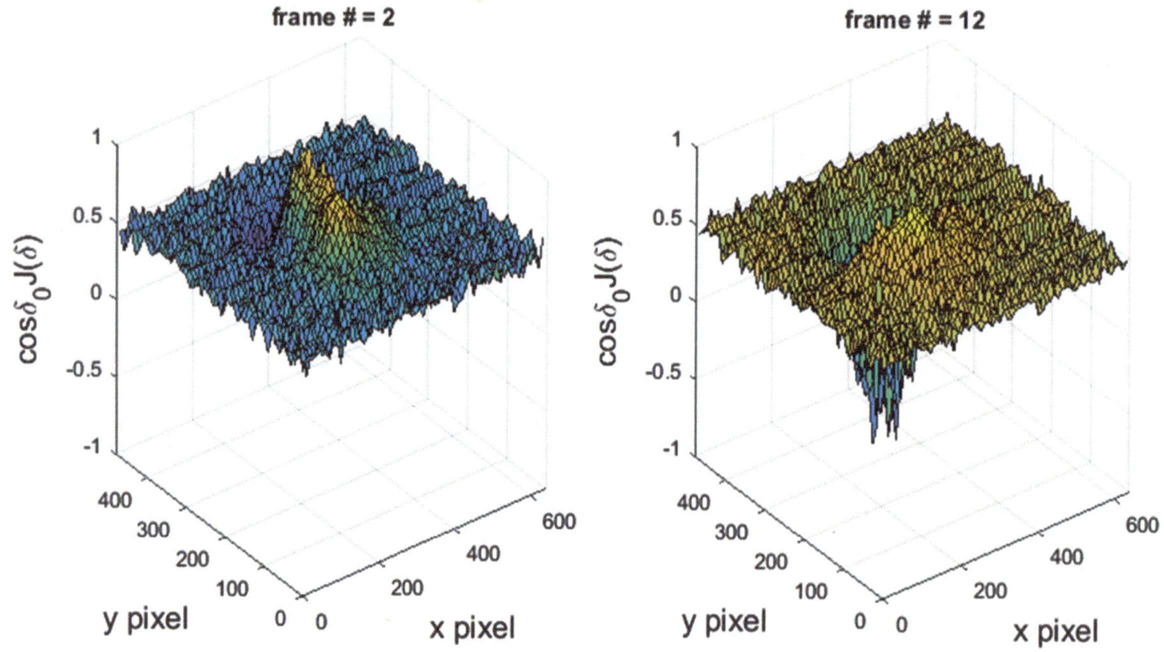


**Fig. 5.6** Interference intensity obtained at frame # 2 and #12

the negative side. It is impossible to determine if this difference comes from the $\cos\delta_o$ or $J_0(\delta)$ term. The frame rate is not synchronized with the oscillation frequency and the interference intensity profile can vary at each time frame. Third, careful comparison of the profiles at frame #2 and 12 indicates that the shapes of the bulge of the two are different; the interference in frame #12 appears to be closer to $x = 0$ than frame #2. This indicates that the way the reference and signal beams overlap each other is different between the two frames.

Figure 5.7a plots the time trends of the peak position of the interference intensity along one row (along a line a parallel to the x-axis near the y center in Fig. 5.6) when the specimen is driven at various frequencies. It is seen that the peak locations fluctuate when the transducer is turned on. The three driving frequencies, i.e., 11.0 kHz, 11.2 kHz and 11.3 kHz show different

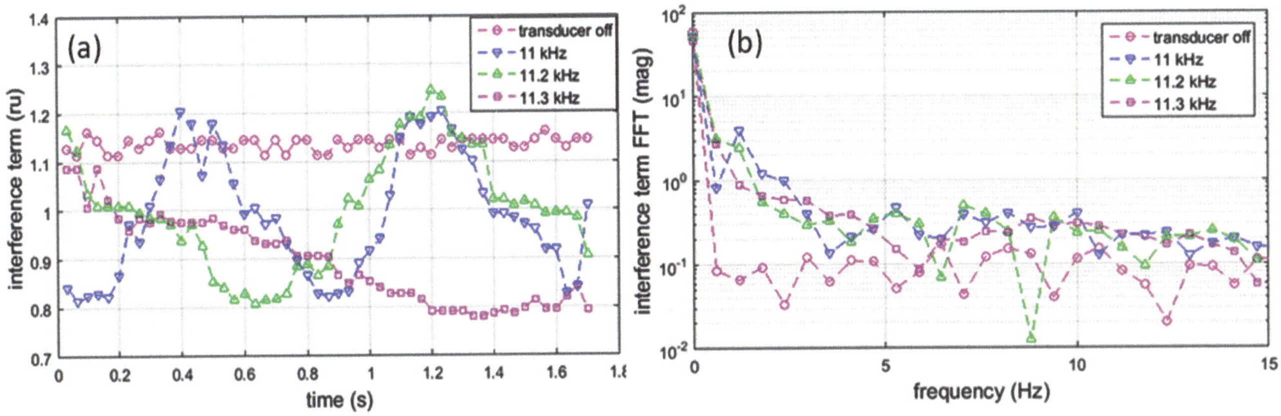


**Fig. 5.7** (**a**) Temporal fluctuation of interference intensity peak location; (**b**) Frequency dependence of (**a**)

trend from one another. The 11 kHz case shows almost sinusoidal fluctuation with a frequency of approximately 1.25 Hz (0.8 s in the period); the 11.2 kHz shows similar trend with a slightly lower frequency; and the 11.3 kHz show much slower change with approximately the same amplitude as the 11.0 kHz and 11.2 kHz cases. There is no correlation between these frequencies and the driving frequencies. It is interpreted that the oscillatory behavior of these time trends is due to a random fluctuation of the phase $\delta_o$. Figure 5.7b plots the frequency dependences of these three time trends. There is no peak in the driving frequency, supporting the interpretation that the oscillatory behaviors seen in Fig. 5.7a is due to a random fluctuation of the phase $\delta_o$. The actual mechanism of the random fluctuation including the reason why the fluctuation is greater when the transducer is turned on is unclear at this time. One possibility is that when turned on the acoustic transducer raises the air temperature and that changes the phase via Eq. (5.6). Our temperature measurement indicates that the air temperature near the specimen can rise of the order of 0.1 °C within 1 s. The corresponding phase change according to Eq. (5.6) is a few percent of the wavelength.

## *5.3.2 Carrier Fringe Configuration*

With the carrier fringe configuration, relative optical path length change causes the dark fringes to shift in a direction perpendicular to the fringes. If such shifts are caused by the acoustic transducer, the digital camera is unable to track the shift because the frame rate is significantly lower than the shift frequency. Consequently, the fringes become blurry, as Fig. 5.4b shows. Mathematically, the fast oscillation decreases the value of $J_0(\delta)$ in Eq. (5.7) as the Bessel function decreases monotonically from 1 to zero as $\delta$ increases from zero towards the first root at 2.4. If the relative optical path length changes slowly, the digital camera can capture the corresponding shift of the carrier fringes without compromising the fringe contrast. By tracking the location of the carrier fringes, we can identify the slow change in the relative optical path length. Mathematically, we can track the shift of $\cos\alpha x$ function in Eq. (5.7) along the $x$-axis.

The top three graphs in Fig. 5.8 plot the location of the carrier fringes as a function of frame number (time). The fringe locations are expressed as deviation from the first frame in the unit of the relative value to the fringe-to-fringe distance (the spatial period). For example, a fringe shift of 0.08 at a certain frame means the fringes of that frame are shifted from the first frame by 8% of the spatial period. These plots correspond to Fig. 5.7a that represents the trend of relative optical path length change observed with the normal incident configuration. As expected, the behavior of the fluctuation is similar to Fig. 5.7a.

The bottom three graphs in Fig. 5.8 are the Fourier spectrum of the spatial intensity variation of the fringe pattern for each frame number; the Fourier spectra of multiple frames out of the 100 frames are superposed in the same graph for each driving frequency. The Fourier spectra are evaluated along a horizontal line crossing the fringes perpendicularly (along row 250 of the fringe image shown in Fig. 5.4). The higher the fringe contrast, the higher the main peak of the Fourier spectrum. Three sets of graphs are selected because of the following features they show. The top graph of (a) exhibits a slow shift of the fringes. In 100 frames (approximately in 3.3 s), the fringes keep shifting in the same direction as much as 10% of the spatial period. However, the lower graph indicates that the Fourier spectrum is practically unchanged. In (b) the fringes remain almost at the same location. For the 3.3 s period, the shift is within a few percent. Yet, the Fourier spectra exhibit some scattered feature at the low frequency side of the spectral peak (between 0.01 and 0.02 1/pixel). The fringe shift in

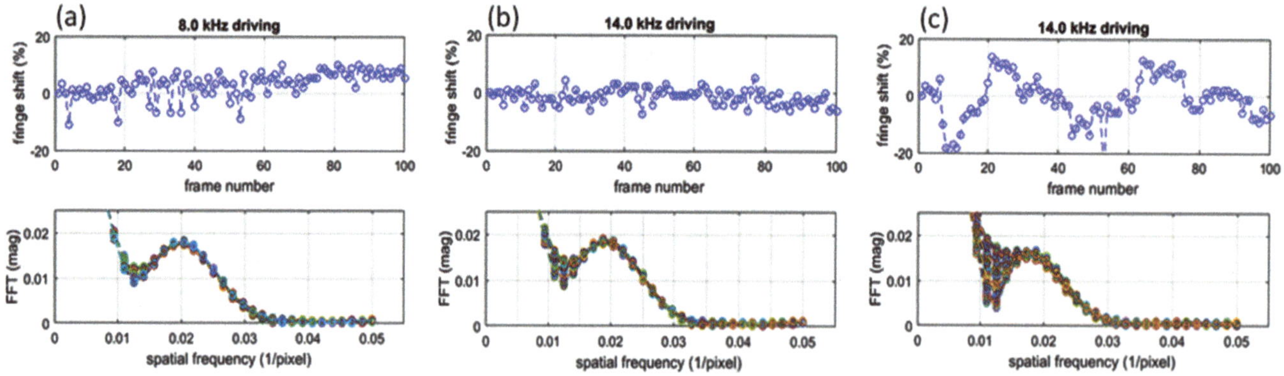


**Fig. 5.8** Three patterns of fringe shift and Fourier spectrum of spatial intensity profile

(c) is the greatest among the three, and the corresponding Fourier spectra exhibit greater feature of the low frequency side scattering. The Fourier spectral peak also shows some changes among different frames.

The above-mentioned features in the three graphs lead to the following observations. From comparison of Figs. 5.8a, b, we can say that a low frequency (slow) fringe shift does not necessarily compromise the shape of the Fourier spectrum. The spectrum peak height and position on the frequency axis are unchanged. Comparison of Fig. 5.8c with the other two indicates that the fringe shift or some accompanying effect can cause the lower frequency side scattering in the Fourier spectrum. Since the comparison of (a) and (b) indicates that a slow fringe shift does not always change the Fourier spectrum shape, it is likely that the reason for the change in the Fourier spectral shape exhibited by Fig. 5.8c is due to some accompany effect than the fringe shift itself.

The fact that a slow fringe shift does not compromise the Fourier spectral shape is favorable for our analysis of the thin-film specimen. As we reported previously [3, 4], we use the spectral peak height to evaluate the blurriness of the carrier fringes. By comparing the peak height with the case when the acoustic oscillation is removed we evaluate the value of $J_0(\delta)$ in Eq. (5.8), and in turn, estimate the oscillation amplitude $d$ from the value of $\delta$ that corresponds to the evaluated $J_0(\delta)$. Hence, it is essential that the spectral peak height is not influenced by the low frequency phase fluctuation. However, the slight change in the spectral peak height and the noisy low frequency side of the spectrum observed in Fig. 5.8c are sources of concern. It is worth while identifying what causes the features observed in the Fourier spectra shown in Fig. 5.8.

To clarify the causes of the spectral features observed in Fig. 5.8, we made a simple numerical model and conducted simulations. The numerical model allows us to shift the fringes on the $x$-axis as a sinusoidal function of time for a given amplitude and frequency (i.e., it can shift the fringes right and left at a given frequency and amplitude), and change the spatial periodicity. The former is to simulate the change in the operation-point phase $\delta_0$, and the latter to simulate a change in the angular alignment of the interferometric arms. The optical intensity profile is in the form of a product of a Gaussian profile and a cosine function. The Gaussian profile is to simulate the laser beam's intensity profile and the cosine function is to simulate the $\cos\alpha x$ term in Eq. (5.7). The model also allows us to vary the spot size of the Gaussian profile. This is to simulate the effect that the signal beam can have a smaller spot size depending on the reflectivity of the specimen surface at the spot where the laser beam is reflected.

Figure 5.9 shows some results of the numerical simulation. Here (a) is when both the fringe shift and the spatial periodicity change is negligible; (b) is when both the fringe shift and spatial periodicity fluctuate; (c) is when only the fringe shift fluctuates in the same fashion as (a); and (d) is when only the spatial periodicity fluctuates in the same fashion as (a). The top plot of each set shows the superposition of optical intensity for 100 frames and the bottom plot shows the corresponding spatial Fourier spectrum. The amplitude of the fringe shift is approximately 15% of the spatial periodicity as the top of Fig. 5.9c indicates. Similarly, the amount of the variation in the spatial periodicity is seen in the top of Fig. 5.9d where the left side of the leftmost peak and the right side of the rightmost peak are expanded to the left and right, respectively; as the spatial periodicity increases the beam profiles expands on the spatial axis. It is clear that the translational shift of the fringes barely affects the spectral shape. This is understandable because purely translational movement of the fringes changes neither the fringe spacing (the spatial periodicity) nor the intensity profile; thus in the frequency domain, the spectrum is unchanged. On the other hand, the angular fluctuation affects the spectrum shape to a certain extent. The bottom plots of Fig. 5.9d shows the spectral peak shows some fluctuation at the high and frequency corner of the peak. This is also understandable as the angular

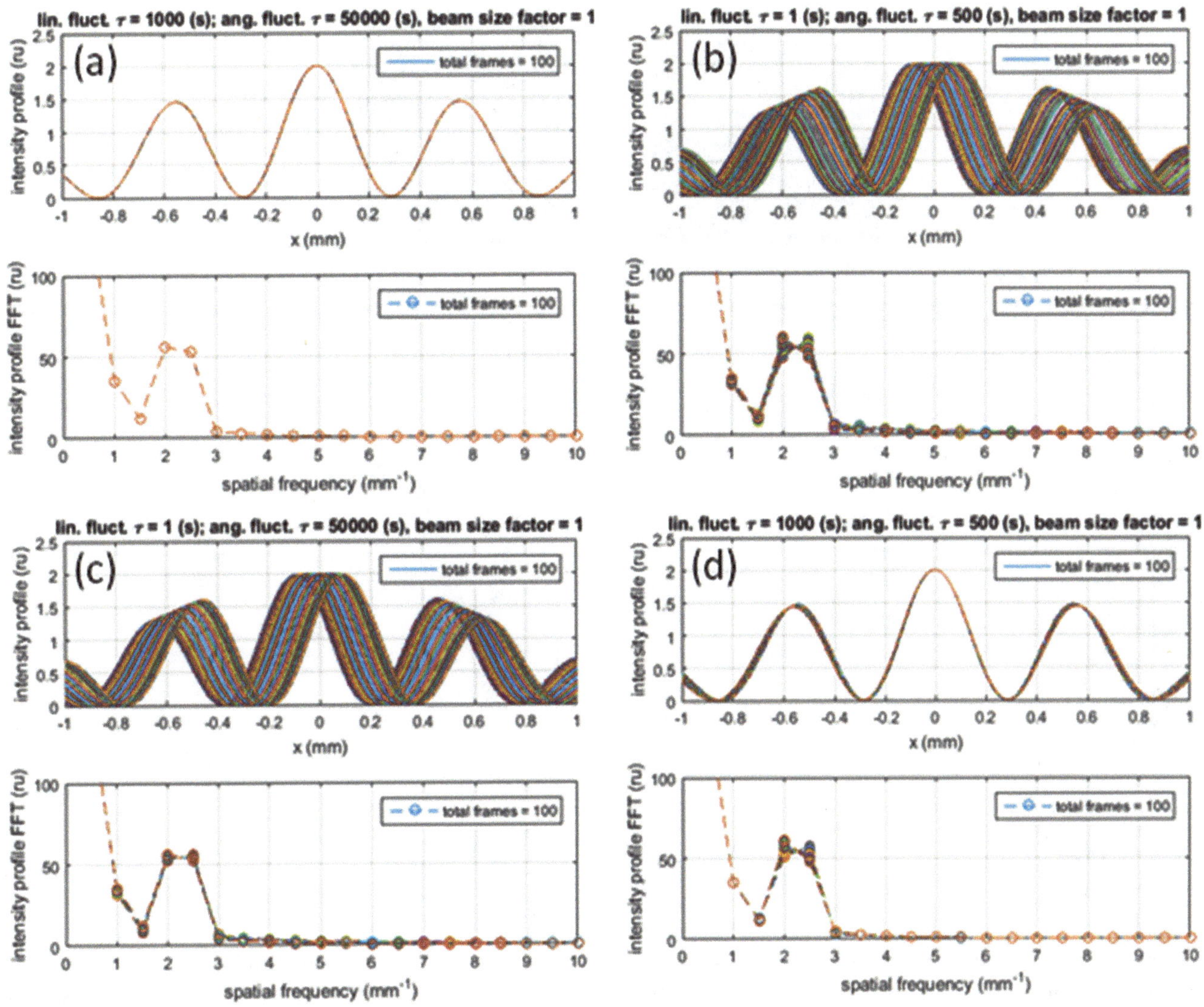


Fig. 5.9 Effect of fringe shifts and angular fluctuation

fluctuation increases the fringe spacing the Fourier spectrum bulges to the low frequency side and as the angular fluctuation decreases the fringe spacing the spectrum bulges in the other direction.

Figure 5.9 indicates neither the fringe shift nor angular fluctuation explains the scattering of the Fourier spectrum on the low frequency foot exhibited by the bottom of Fig. 5.8c. We have found that the change in the size of the combined (interfering) laser beams and its relative motion account for this effect. Figure 5.10a shows the actual image of the data shown in Fig. 5.8c at frame #1 and # 100. Apparently, the brightest spot of the beam is shifted toward left from frame #1 to #100. Figure 5.10b, c are numerical results in which we move the intensity profile (more specifically we shift the Gaussian profile) at 1 Hz back and forth with no fringe shift or angular fluctuation. The amplitude of this motion is set to be five times of the fringe shift used in Fig. 5.9b, c. Here Fig. 5.10b uses the same beam spot size as Figs. 5.9 and 5.10c uses 40% narrower beam size. It is clearly seen that when we move the actual beam with the same beam size, the numerical simulation does not reproduce the low frequency side scattering observed in Fig. 5.8c, whereas it clearly exhibits the effect if the beam size is reduced to 40%. This is reasonable as a smaller beam size has more influence on the Fourier spectrum of the intensity profile.

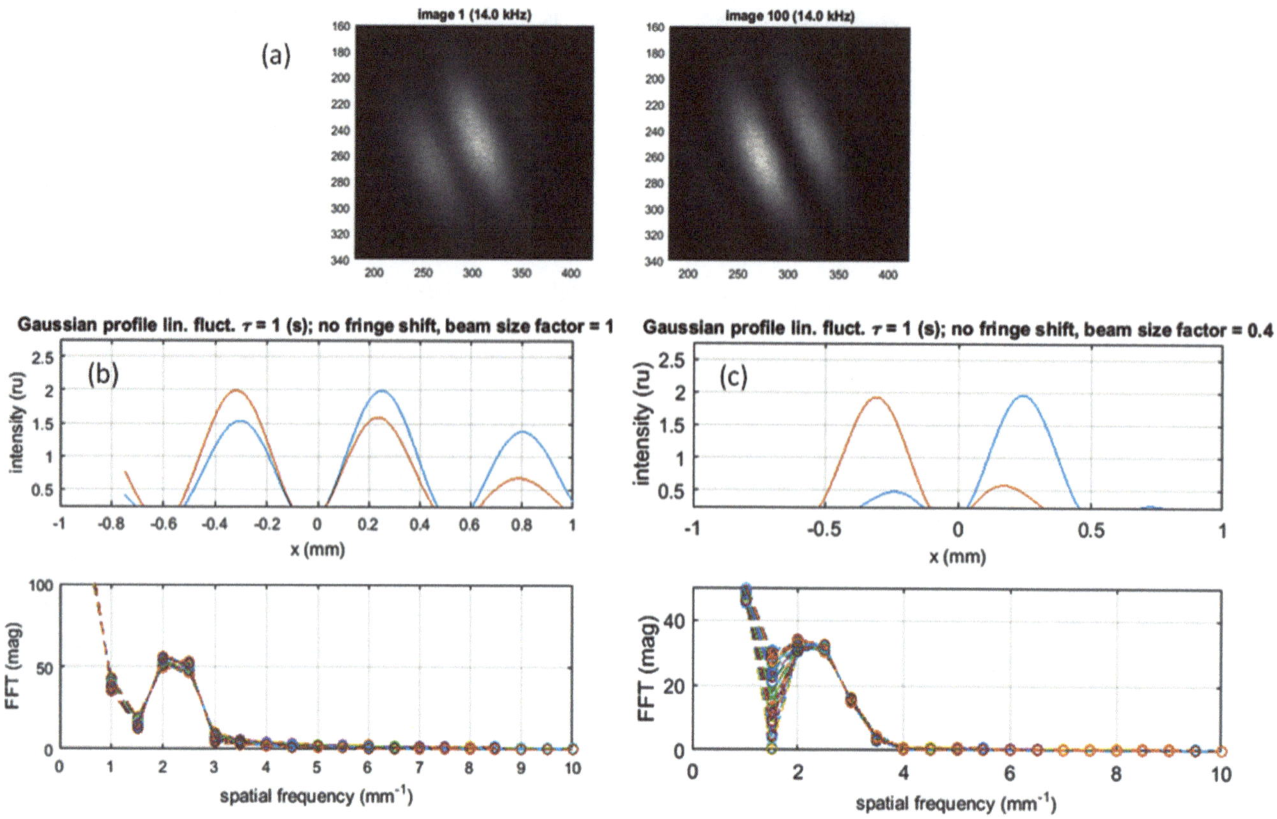


Fig. 5.10 Effect of beam size and motion

## 5.4 Summary

The influence of the operation-point (initial) phase noise on the accuracy of Amplitude-Fluctuation Electronic Speckle-Pattern Interferometry is discussed. It has been demonstrated that the use of carrier fringes and analysis in the spatial frequency domain reduces the noise associated with a slow fluctuation of the operation-point phase. It has been found that movement of the interferometric beams can compromise the frequency domain analysis as it can vary the spectral peak height.

**Acknowledgement** This work was supported by the Louisiana Board of Regents Pilot-fund grant LEQSF(2016-17)-RD-C-13, Southeastern Louisiana University STAR grant, the National Research Foundation of Korea (NRF) grant funded by the Korea government (MSIP) and NRF-2013M2A2A9043274, NRF-2011-220-D00002.

## References

1. Huang, C., Ma, C.: Vibration characteristics for piezoelectric cylinders using amplitude- fluctuation electronic speckle pattern interferometry. AIAA J. **36**, 2262–2268 (1988)
2. Wang, W.C., Hwang, C.H.: The development and applications of Amplitude Fluctuation Electronic Speckle Pattern Interferometry method. In: Kounadis, A.N., Gdoutos, E.E. (eds.) Recent Advances in Mechanics, pp. 343–358. Springer, New York (2011)
3. Yoshida, S., Adhikari, S., Gomi, K., Shrestha, R., Huggett, D., Miyasaka, C., Park, I.K.: Opto-acoustic technique to evaluate adhesion strength of thin-film systems. AIP Adv. **2**, 022126:1–022126:7 (2012)
4. Yoshida, S., Didie, D.R., Sasaki, T., Park, H.S., Park, I.K., Gurney, D.: Opto-acoustic method for characterization of the thin-film adhesion. Appl. Sci. **6**, 163 (2016). doi:10.3390/app6060 163
5. Owens, J.C.: Optical refractive index of air: dependence on pressure, temperature and composition. Appl. Opt. **6**(1), 51–59 (1967)

**Sanichiro Yoshida** received his undergraduate and doctoral degrees from Keio University. Currently he is with Department of Chemistry and Physics at Southeastern Louisiana University. His research interest includes field theory, optical interferometric and opto-acoustic methods for experimental mechanics and condense matter physics.

# Chapter 6
# Evaluating Path of Stress Triaxiality to Fracture of Thin Steel Sheet Using Stereovision

**D. Kanazawa, S. Chinzei, Y. Zhang, K. Ushijima, J. Naito, and S. Yoneyama**

**Abstract** A stereovision technique based on digital image correlation is applied to the evaluation of the stress triaxiality and fracture strain of thin steel sheet. A tensile testing specimen with notches made of high strength steel sheet is loaded and the surface displacements are measured from both sides of the specimen surface using two stereovision systems. Not only the in-plane strains but the through-thickness strains are evaluated from the measurement results of the displacements on the both surfaces of the specimen. The variation of the stress triaxiality at an evaluation point is evaluated from the measured strains. The fracture strain is also evaluated from the strain measurement results. Experimental results show that the stress triaxiality and the fracture strain of thin steel sheet can be evaluated by the surface strain measurement. The results can be utilized for simulating deformation and predicting fracture of a component made of thin steel sheet.

**Keywords** Stereovision • Digital Image Correlation • Stress Triaxiality • Fracture Strain • Through-thickness Strain

## 6.1 Introduction

Thin and high strength steel sheets are widely used for automobile body frames to realize lightweight and strong bodies. In order to design reliable body frames, the predictions of deformation and fracture phenomenon of the thin steel sheet at the time of a crash have become important. The fracture of metal materials occurs after the necking developing. It is known that the onset of the necking of the materials depends on the deformation and the stress state of the materials [1]. Similarly, the fracture strain depends on the stress state of the materials. The stress triaxiality is one of the parameters that show the stress state of materials. Therefore, the evaluation of the stress triaxiality as well as the fracture strain has become essential for predicting the fracture.

Ostuka et al. [2] evaluated the fracture strains and the stress triaxialities at the time of the fracture experimentally with Bridgman's formula and revealed that the fracture strains were dependent on the stress triaxiality. On the other hand, Ma et al. [4, 5] used FEM analysis for evaluating the stress triaxiality and reported that the value of the stress triaxiality varied with the deformation of materials. Therefore, the variation of the stress triaxiality with deformation as well as the stress triaxiality at fracture should be evaluated as one of the material properties for the precise prediction of the fracture phenomena. Bai and Wierzbicki [6] proposed a method for evaluating the stress triaxiality from the increments of principal plastic strains. However, it is considered that strain increments obtained by experiment are suffered from experimental errors because measured displacements have a lot of errors in their derivatives.

In this study, the variation of the stress triaxiality and the fracture strain of thin steel plates are evaluated experimentally. A stereovision technique based on digital image correlation [7] is used for this purpose. A tensile testing specimen with notches made of high strength steel sheet is loaded and the surface displacements are measured from both sides of the

D. Kanazawa • Y. Zhang • S. Yoneyama (✉)
Department of Mechanical Engineering, Aoyama Gakuin University, 5-10-1 Fuchinobe, Sagamihara, 252-5258, Japan
e-mail: yoneyama@me.aoyama.ac.jp

S. Chinzei • J. Naito
Mechanical Engineering Research Laboratory, Kobe Steel, LTD, 1-5-5 Takatsukadai, Nishi, Kobe, 651-2271, Japan

K. Ushijima
Department of Mechanical Engineering, Tokyo University of Science, 6-3-1 Niijyuku, Katsushika, Tokyo, 125-8585, Japan



L. Lamberti et al. (eds.), *Advancement of Optical Methods in Experimental Mechanics, Volume 3*,
Conference Proceedings of the Society for Experimental Mechanics Series, DOI 10.1007/978-3-319-63028-1_6

specimen surface using two stereovision systems. Not only the in-plane strains but also the through-thickness strains are evaluated from the measurement results of the displacements on the both surfaces of the specimen. The variation of the stress triaxiality at an evaluation point is evaluated from the measured strains. The fracture strain is also evaluated from the strain measurement results. Experimental results show that the stress triaxiality and the fracture strain of thin steel sheet can be evaluated by the surface strain measurement. The results can be utilized for simulating deformation and predicting fracture of a component made of thin steel sheet.

## 6.2 Experimental Procedure

The shape and the size of the specimen used in this study are shown in Fig. 6.1. Three types of high strength steel, the tensile strengths of more than 1470 MPa (TS1470MPa), 980 MPa (TS980MPa), and 590 MPa (TS590MPa) are used as the test materials. Notched thin sheet specimens made from above materials are loaded by a tensile testing machine, and the deformation of the both surfaces of the specimen are measured using two stereo digital image correlation systems, as show in Fig. 6.2. For digital image correlation, the specimen is painted with black ink on the surface, and a white dot pattern

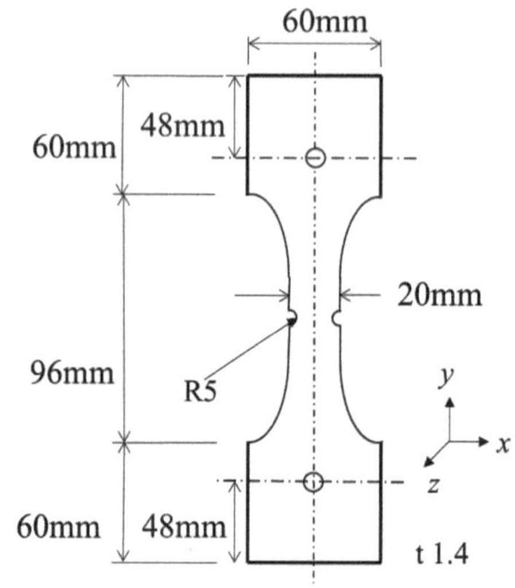


**Fig. 6.1** Shape of specimen

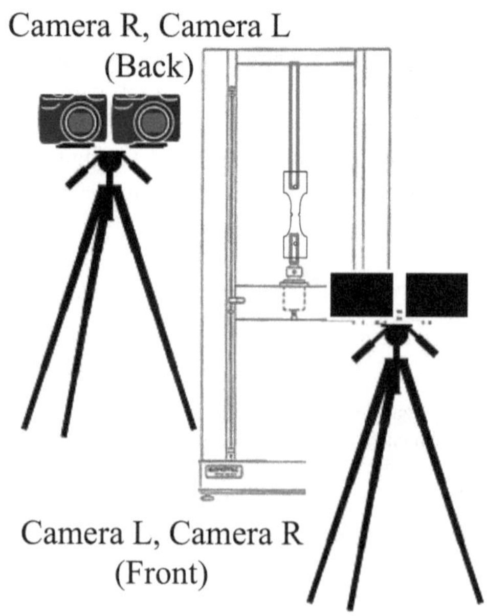


**Fig. 6.2** Experimental setup

is overlaid by spray painting, creating a speckle-like pattern. The size of the random pattern is selected to oversample the intensity pattern using several sensors for accurate measurement. In this study, each random pattern is oversampled by 10–40 pixels. The variations in the speckle-like pattern on the specimen are observed by four monochromatic CCD cameras (2048 × 2048 pixels × 8 bits) equipped with a lens with a focal length of 105 mm. A bicubic interpolation method is used for obtaining the continuous speckle pattern. The data processing is implemented using software developed by one author.

## 6.3 Results

Figure 6.3 shows the example set of the specimen surface observed by a stereovision. Using the principle of the stereovision, three-dimensional surface position before and after the load are determined. Then, the three-dimensional displacements are obtained as shown in Fig. 6.4. The strain distributions are computed from the displacement distributions of the both surfaces. Here, it is noted that the logarithmic strains are employed as the strain measure. Figure 6.5 shows the example set of the logarithmic principal strain distributions for the TS980MPa specimen under the load of $P = 14.4$ kN. Note that the third principal strain $\varepsilon_3$ represents the through-thickness strain. As shown in this figure, it is observed that the strains concentrate at the middle of the specimen and the reduction of the thickness is obvious.

From the measured strains, the variations of the stress triaxiality at the center of the specimen are evaluated. The stress triaxialites obtained by various procedures are shown in Fig. 6.6. The symbol $\eta_{e2}$ represents the stress triaxiality evaluated form the in-plane principal strain increments, $\eta_{e3}$ gives the value evaluated from the in-plane principal strain increments as well as the through-thickness principal strains. As mentioned in Introduction, the experimentally obtained strain increments are suffered from the greater measurement errors than the strains. Therefore, the stress triaxialities are also evaluated from the stresses that are computed from the strains using the total strain theory [8]. The symbol $\eta_{s2}$ is the stress triaxiality obtained from the in-plane stresses and the symbol $\eta_{se}$ expresses the value computed from the in-plane and through-thickness stresses. It is observed from this figure that the stress triaxialities change depending on the deformation of the specimens and tend to gradually approach the nominal value $\eta = 0.577$. The smooth variations are observed for the values obtained from the stresses whereas the values obtained from the strain increments exhibit not smooth variation. Further studies are required for validating these results.

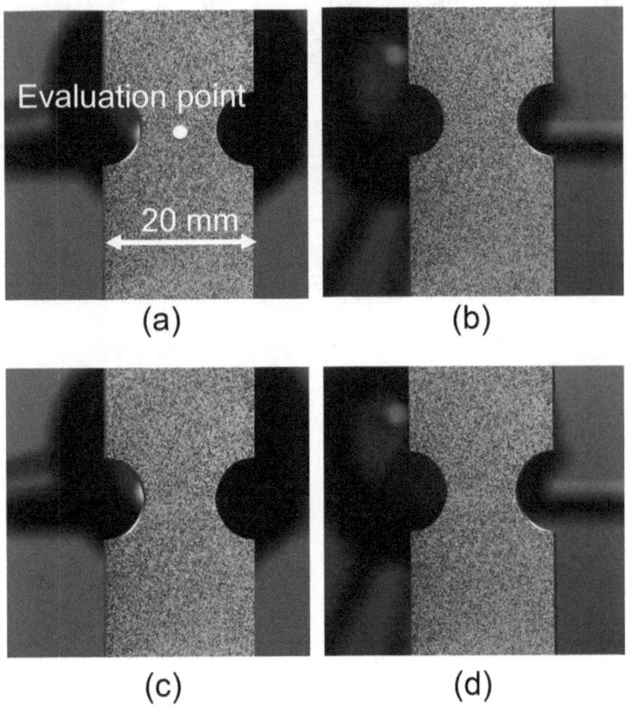


**Fig. 6.3** Example set of stereo images of front surface for TS980MPa specimen: (**a**) left image before load; (**b**) right image before load (**c**) left image after load ($P = 14.4$ KN); (**d**) right image after load ($P = 14.4$ KN)

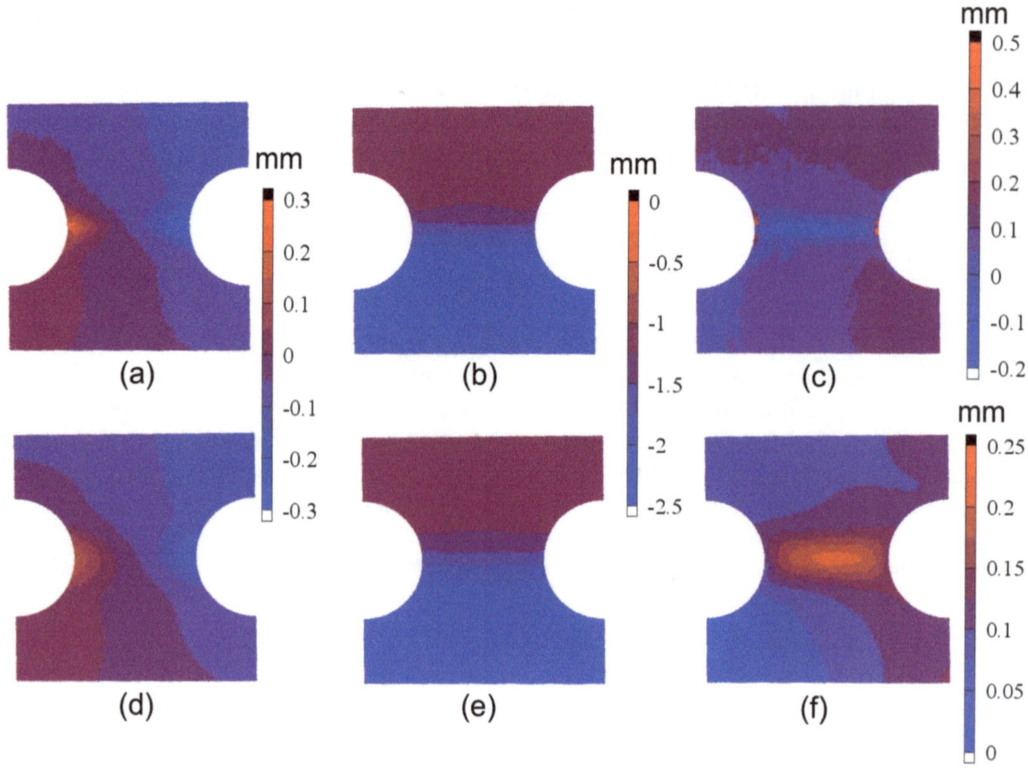


**Fig. 6.4** Example set of displacement distributions for TS980MP specimen ($P$ = 14.4 KN): (**a**) $u_x$ on front surface; (**b**) $u_y$ on front surface; (**c**) $u_z$ on front surface; (**d**) $u_x$ on back surface; (**e**) $u_y$ on back surface; (**f**) $u_z$ on back surface

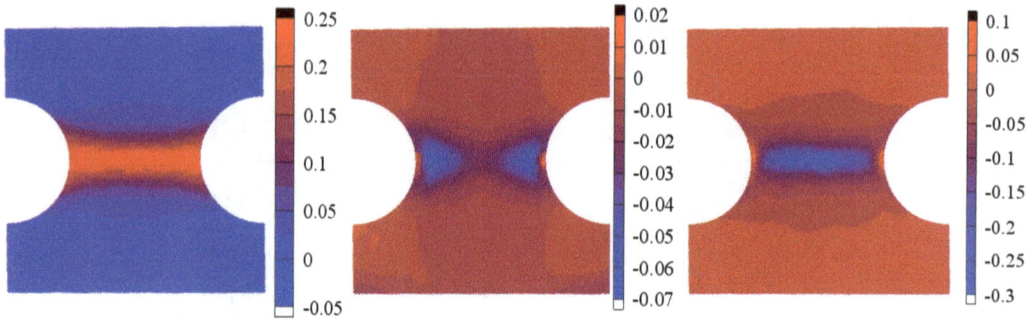


**Fig. 6.5** Example set of strain distributions for TS980MP specimen ($P$ = 14.4 KN): (**a**) $\varepsilon_1$; (**b**) $\varepsilon_2$; (**c**) $\varepsilon_3$

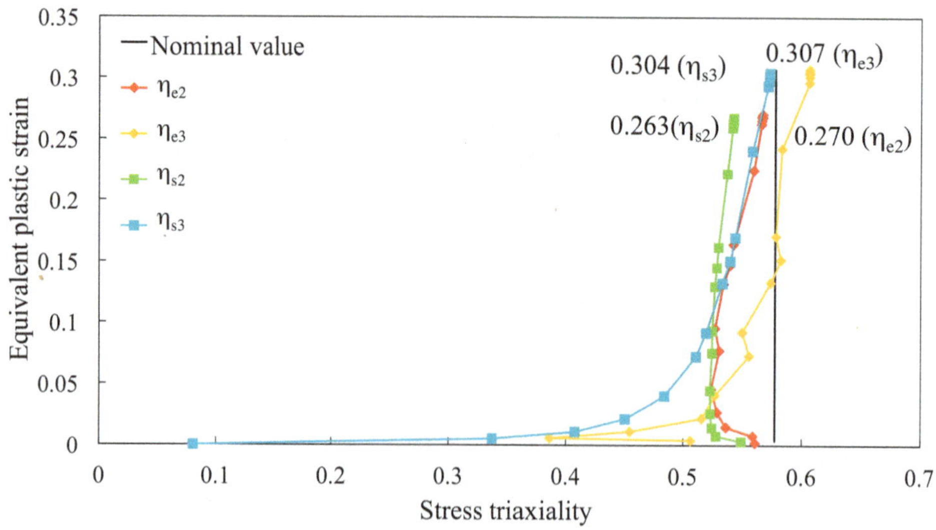


**Fig. 6.6** Variations of stress triaxiality obtained using various procedures (TS980MPa)

## 6.4 Conclusion

A stereovision technique based on digital image correlation is applied to the evaluation of the stress triaxiality and fracture strain of thin steel sheet. A tensile testing specimen with notches made of high strength steel sheet is loaded and the surface displacements are measured from both sides of the specimen surface using two stereovision systems. Not only the in-plane strains but the through-thickness strains are evaluated from the measurement results of the displacements on the both surfaces of the specimen. The variation of the stress triaxiality at an evaluation point is evaluated from the measured strains. The fracture strain is also evaluated from the strain measurement results. Experimental results show that the stress triaxiality and the fracture strain of thin steel sheet can be evaluated by the surface strain measurement. The results can be utilized for simulating deformation and predicting fracture of a component made of thin steel sheet.

## References

1. Effelsberg, J., Haufe, A., Feucht, M., Neukamm, M., Du Bois, P.: On parameter identification for the GISSMO damage model. In: 12th International LS-DYNA Users Conference, pp. 1–12 (2012)
2. Ostuka, A., Miyata, T., Nishimura, S., Kimura, M., Mabuchi, M.: Effect of stress triaxiality on ductile fracture initiation of low strength steels. J. JSMS. **29**(322), 717–723 (1979)
3. Bridgman, J., .W.: Studies in Large Plastic Flow and Fracture. Harvard University Press, Cambridge (1952)
4. Ma, N., et al.: Study on fracture prediction on advanced high strength steels. Trans. JASE. **47**(2), 603–608 (2016)
5. Sato, K., et al.: Study on fracture prediction on advanced high strength steels. Trans. JASE. **45**(6), 1099–1104 (2014)
6. Bai, Y., Wierzbicki, T.: Forming severity concept for predicting sheet necking under complex loading histories. Int. J. Mech. Sci. **50**, 1012–1022 (2008)
7. Yonayama, S.: Measurement method for displacement and strain on two- and three-dimensional surface by using digital image correlation. Journal of the JSTP. **55**(646), 979–983 (2014)
8. Kunio, T.: Introduction to Solid Mechanics. Baifukan, Tokyo (1977)

**S. Yoneyama** 1995 BS Aoyama Gakuin University; 1997 MS Aoyama Gakuin University; 2000 PhD Tokyo Institute of Technology; 2000 Research Associate Wakayama University; 2004 Assistant Professor Osaka Prefecture University; 2007 Associate Professor Aoyama Gakuin University; 2015 Professor Aoyama Gakuin University.

# Chapter 7
# Studying with a Full-Field Measurement Technique the Local Response of Asphalt Specimens Subjected to Freeze-Thaw Cycles

**M. C. Teguedi, B. Blaysat, E. Toussaint, S. Moreira, S. Liandrat, and M. Grédiac**

**Abstract** Asphalt is a strongly heterogeneous material, whose global thermo-mechanical behavior is generally studied at a large spatial scale. Observing the local strain fields which occur between aggregates and mastic (i.e., asphalt binder + filler) is however crucial to understand the phenomena that influence the global response of such materials, and then better design them. In this study, full-field measurements were used to investigate the difference in behavior between various types of asphalt containing Recycled Asphalt Pavement (RAP). We focused here on their response when they were subjected to freeze-thaw cycles. The local contraction/expansion of the specimens subjected to various cooling/heating phases was observed with the grid method. The strain distribution has been found to be very heterogeneous because of the difference in coefficient of thermal expansion between aggregates and mastic. The influence of the percentage of RAP on the global response of the specimen is observed and discussed.

**Keywords** Recycled Asphalt • Thermal contraction/expansion • Strain field • Freeze-thaw • Grid method • Digital image correlation

In cold regions, pavements suffer from the thermal cracking caused by the contraction and expansion of asphalt under temperature changes. The thermal behavior of asphalt at the macroscopic scale has been widely studied using testing devices such as the Asphalt Thermal Cracking Analyzer (ATCA) [1] and the Asphalt Concrete Cracking Device (ACCD) [2]. However, since low temperature damages are initiated in the internal structure of asphalt mixtures, understanding the thermal response of asphalt components (bitumen and aggregates) is of great practical importance to have a better understanding of the overall thermal response of pavements. The major problem is however to measure the local thermal response of these materials.

Full-field measurement techniques, which have now widely spread in experimental mechanics, are useful to have a better understanding of the asphalt behavior under thermal loadings. In recent years, these methods have become major tools to inspect and characterize the mechanical response of such materials. The main full-field measurement methods that are used by the pavement community are the digital image correlation (DIC) [3–5] and more recently the grid method (GM) [6, 7]. This last technique was used here because it exhibits a good compromise between spatial resolution and measurement resolution [8], which is crucial here, the material under investigation being highly heterogeneous. Till now, the application of such measurement techniques for the characterization of asphalt was limited to their mechanical response, and few emphasis has been placed on the investigation of their thermal response.

In this study, four Hot Mixtures Asphalt (HMA) specimens with 0%, 20%, 40% and 100% of RAP (Recycled Asphalt Pavement) content were considered. These materials were used in a previous study, which was devoted to the characterization of the effect of RAP on the local mechanical behavior of recycled asphalt pavements [6]. These RAP materials are composed of granite, basalt and gneiss. The virgin materials are constituted from limestone aggregates and a virgin bituminous binder. Freeze-thaw tests were carried out on these materials. They were performed in a climate chamber at a temperature range of [−10 °C 20 °C]. At the global scale, the specimens exhibited an anisotropic behavior along both the vertical and horizontal directions. The comparison of the global strain-temperature curves showed that the inclusion of 100% RAP resulted in an increase of the Coefficient of Thermal Contraction (CTC) of the specimen. Typical examples of the displacement fields

M.C. Teguedi • B. Blaysat • E. Toussaint • M. Grédiac (✉)
Université Clermont Auvergne, Institut Pascal, BP 10448, 63000, Clermont-Ferrand, France

CNRS, UMR 6602, Institut Pascal, 63171, Aubière, France
e-mail: michel.grediac@univ-bpclermont.fr

S. Moreira • S. Liandrat
Département Laboratoire de Clermont-Ferrand (CEREMA, DterCE), Clermont-Ferrand, France



L. Lamberti et al. (eds.), *Advancement of Optical Methods in Experimental Mechanics, Volume 3*,
Conference Proceedings of the Society for Experimental Mechanics Series, DOI 10.1007/978-3-319-63028-1_7

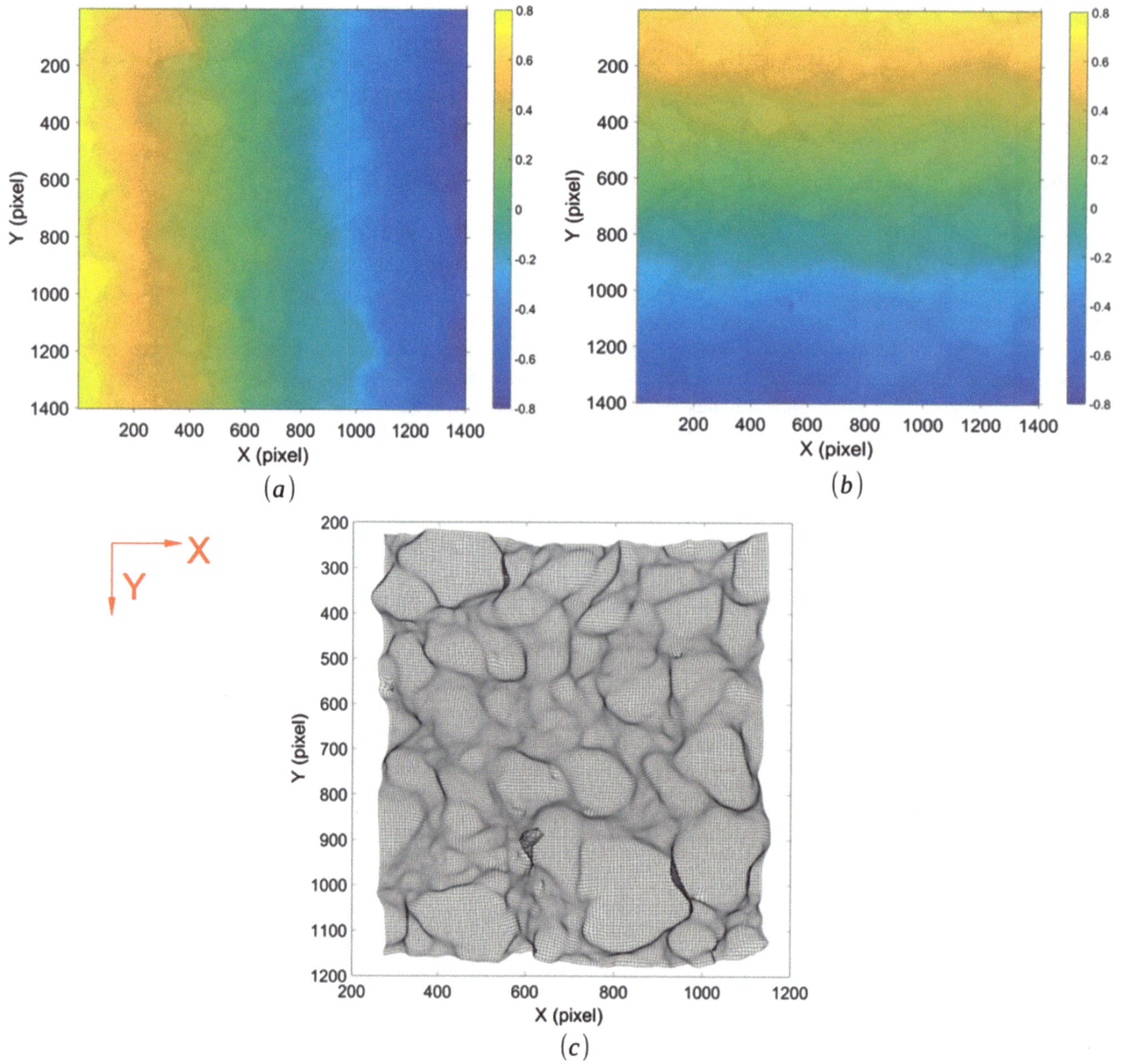


**Fig. 7.1** Typical displacement fields for the 100% RAP specimen at T = −10 °C (in pixels, 1 pixel = 40 μm). (**a**) $U_{xx}$, 20% *RAP*. (**b**) $U_{yy}$, 20% *RAP*. (**c**) Magnified deformed mesh

obtained for the 100% RAP specimen at T = −10 °C are shown in Fig. 7.1. These displacements correspond to a difference in temperature equal to ΔT = −30 °C. The displacement fields presented in Fig. 7.1a, b clearly illustrate the global contraction of the specimen along the vertical and horizontal direction. From these results, it is visible that the specimen shrinks toward its center. Another representation of the displacement fields is proposed in Fig. 7.1c. In this figure, the displacement fields are used to deform a regular mesh to help the reader figure out the deformation of the specimen. The pitch of the mesh is equal to 10 pixels and the displacement multiplied by 350 to give a clearer idea on the deformation of the specimen. This presentation is expressed in the deformed configuration. The overall contraction of the specimen towards its center is clearly visible. This contraction is mainly sustained by the mastic. Strain maps which are deduced from these displacement maps then enable us to characterize, among others, the in-situ coefficient of thermal expansion of the binder.

Finally, it can be concluded that the paper brings new information on the thermal contraction/expansion of asphalt materials at length scales ranging from binder to the mixture scales. In particular, it was possible to quantify the influence of the RAP on this response.

Full details on these experiments can be found in [9].

## References

1. Stimilli, A., Canestrari, F., Teymourpour, P., Bahia, H.U.: Low-temperature mechanics of hot recycled mixtures through Asphalt Thermal Cracking Analyzer (ATCA). Constr. Build. Mater. **84**, 54–65 (2015)
2. Akentuna, M., Kim, S.S., Nazzal, M., Abbas, A.R., Arefin, M.S.: Study of the thermal stress development of asphalt mixtures using the Asphalt Concrete Cracking Device (ACCD). Constr. Build. Mater. **114**, 416–422 (2016)
3. Romeo, E.: Two-dimensional digital image correlation for asphalt mixture characterisation: interest and limitations. Road Mater. Pavement Des. **14**(4), 747–763 (2013)
4. Ktari, R., Millien, A., Fouchal, F., Pop, I.O., Petit, C.: Pavement interface damage behavior in tension monotonic loading. Constr. Build. Mater. **106**, 430–442 (2016)
5. Hill, B., Buttlar, W.G.: Evaluation of polymer modification in asphalt mixtures through digital image correlation and performance space diagrams. Constr. Build. Mater. **122**, 667–673 (2016)
6. Teguedi, M.C., Blaysat, B., Toussaint, E., Moreira, S., Liandrat, S., Grédiac, M.: Applying a full-field measurement technique for studying the local deformation in reclaimed asphalt pavements. Constr. Build. Mater. **121**, 547–558 (2016)
7. Teguedi, M.C., Toussaint, E., Blaysat, B., Moreira, S., Liandrat, S., Grédiac, M.: Studying the influence of the reclaimed asphalt pavement (RAP) on local deformation properties of asphalt mixtures. In: Zhu, Y., Zehnder, A.T. (eds.) Experimental and Applied Mechanics, pp. 159–163. Springer International Publishing (2017)
8. Grédiac, M., Blaysat, B., Sur, F.: A critical comparison of some metrological parameters characterizing local digital image correlation and grid method. In revision
9. Teguedi, M.C., Toussaint, E., Blaysat, B., Moreira, S., Liandrat, S., Grédiac, M.: Towards the local expansion and contraction measurement of recycled asphalt pavements (RAP) exposed to freeze-thaw cycles. Constr. Build. Mater. **154**, 438–450 (2017)

**M. Grédiac** 1989: PhD, University of Lyon, France; 1991–1997: Assistant professor, University of St-Etienne, France; 1997-date: professor, Université Clermont Auvergne, Clermont-Ferrand, France.

# Chapter 8
# Mechanical Shape Correlation: A Novel Integrated Digital Image Correlation Approach

**S.M. Kleinendorst, J.P.M. Hoefnagels, and M.G.D. Geers**

**Abstract** Mechanical Shape Correlation (MSC) is a novel integrated digital image correlation technique, used to determine the optimal set of constitutive parameters to describe the experimentally observed mechanical behavior of a test specimen, based on digital images taken during the experiment. In contrast to regular digital image correlation techniques, where grayscale speckle patterns are correlated, the images used in MSC are projections of the sample contour. This enables the analysis of experiments for which this was previously not possible, because of restrictions due to the speckle pattern. For example, analysis becomes impossible if parts of the specimen move or rotate out of view as a result of complex and three-dimensional deformations and if the speckle pattern degrades due to large deformations. When correlating on the sample outline, these problems are overcome. However, it is necessary that the outline is large with respect to the structure volume and that its shape changes significantly upon deformation, to ensure sufficient sensitivity of the images to the model parameters. Virtual experiments concerning stretchable electronic interconnects, which because of their slender wire-like structure satisfy the conditions for MSC, are executed and yield accurate results in the objective model parameters. This is a promising result for the use of the MSC method for tests with stretchable electronics and other (micromechanical) experiments in general.

**Keywords** Mechanical shape correlation • Digital image correlation • Integrated digital image correlation • Stretchable electronics • Parameter identification

## 8.1 Introduction

Mechanical Shape Correlation (MSC) is a novel technique, based on integrated digital image correlation (I-DIC) [1, 2], where a finite element (FE) model is coupled to the image correlation procedure. In such a method the constitutive parameters of the FE model are the unknowns in the correlation procedure, with the objective to obtain a good set of model parameters that describe the experimentally observed behavior of the test specimen correctly. In contrast to images of a speckle pattern applied to the test specimen, as usually used in DIC, in MSC the images used for correlation are projections of the deforming sample shape. This is beneficial in cases where complex three-dimensional deformations occur, such that parts of the specimen rotate out-of-view and other parts rotate into view, or if pattern application is difficult, such as on microscale samples [3]. Usually in DIC approaches correlation is limited to in-plane deformations, or in case of Quasi-3D DIC (also referred to as Digital Height Correlation, DHC) [4–7] or stereo-DIC [8, 9] it is also possible to track the out-of-plane deformation of the surface. However, this surface is required to stay in view and not to move or rotate out of view. By correlating the outline of the specimen this restriction is relaxed and full three-dimensional movement of the sample is allowed. However, for the MSC method the assumption is that the boundary area is large with respect to the volume of the structure and that it changes shape significantly upon deformation, such that deformations are reflected in the specimen edge.

A schematic depiction of the method can be seen in Fig. 8.1. During an experiment pictures are taken of the specimen. These images are processed in order to obtain a projection. Also a numerical simulation is executed and similar projections are made. The projections from the experiment and the simulation are compared. If they do not correlate, the model parameters, which are the unknown in the correlation procedure, are updated and a new simulation is performed. This iterative operation is repeated until convergence is reached and the correct set of model parameters to portray the experimental behavior of the structure is achieved.

S.M. Kleinendorst (✉) • J.P.M. Hoefnagels • M.G.D. Geers
Department of Mechanical Engineering, Eindhoven University of Technology, Gemini-Zuid 4.122, 5600MB, Eindhoven, The Netherlands;
e-mail: s.m.kleinendorst@tue.nl; j.p.m.hoefnagels@tue.nl



L. Lamberti et al. (eds.), *Advancement of Optical Methods in Experimental Mechanics, Volume 3*,
Conference Proceedings of the Society for Experimental Mechanics Series, DOI 10.1007/978-3-319-63028-1_8

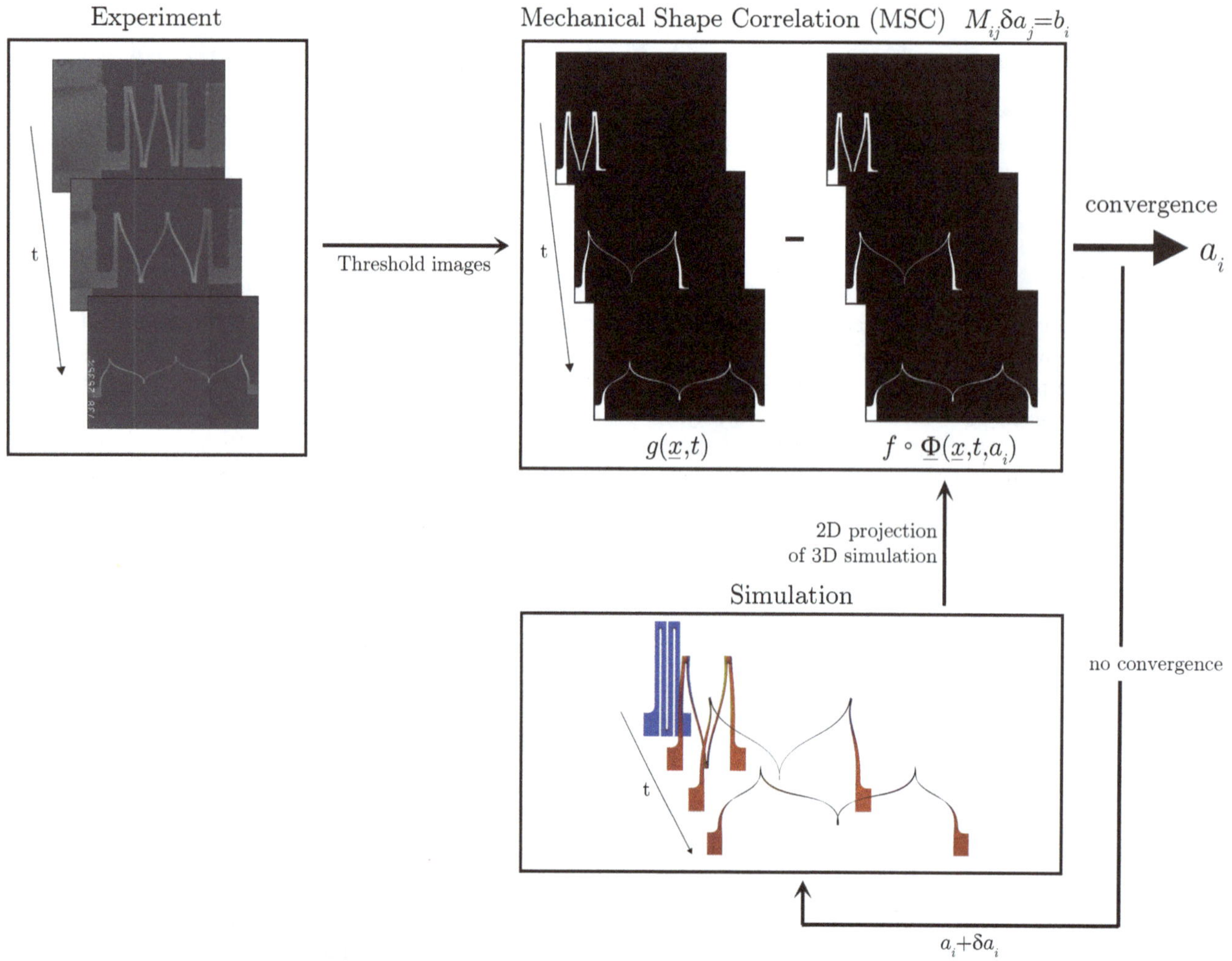


**Fig. 8.1** Schematic representation of the mechanical shape correlation (MSC) method to identify model parameters $a_i$ from an experiment. Simulations supply the images $f$ to which the experiment is compared

The method is demonstrated by means of virtual experiments, dedicated to stretchable electronic interconnects. These interconnects are wire-like structures and hence their outline is large with respect to the volume of the structure. In this case the interconnects are not glued to a stretchable substrate, but they are free-standing and hence free to deform three-dimensionally, making it difficult to analyze their mechanical behavior with existing techniques. Therefore, these structures make an interesting test case for the MSC technique. The deformation of the interconnects can globally be split in two main modes: torsion of beams and double (S-shaped) bending of beams. These two modes are treated separately in the virtual experiments in this work.

The outline of the paper is as follows. In Sect. 8.2 the MSC algorithm is introduced and compared to the integrated digital image correlation algorithm, highlighting the differences. Next the image type used for Mechanical Shape Correlation, i.e., the specimen projections, are explained and the important steps in the formation of these images are stepped through. In Sect. 8.4 two virtual experiment concerning the main deformation modes in the stretchable electronic interconnect structures are presented. In the last section conclusions are drawn.

## 8.2 Algorithm

The algorithm for Mechanical Shape Correlation is based on Integrated Digital Image Correlation (I-DIC) methods, see e.g. [1]. In I-DIC grayscale images of the sample, containing a speckle pattern, before and after deformation are compared, based on the optical flow relation, which means that material points are assumed to have the same gray value before and after deformation. A Finite Element (FE) simulation is performed which represents the experiment, in which some model parameters of interest are the unknowns in the correlation procedure. The resulting displacement field from this simulation is used to back-deform the deformed images, such that if the displacement field is calculated correctly the back-deformed image matches the undeformed, reference, image. The difference between these images is denoted as the *residual*, which naturally approaches zero if a good correlation is obtained. If this is not the case, the unknown model parameters are updated and the procedure is repeated.

The difference in MSC is that the displacement field resulting from the FE simulation is not used to create back-deformed images, but projections, i.e., images, are generated from the model itself at different time increments. This image type is introduced in the next section. These FE images are correlated to similar projections that are made from the experimental images at the same time steps. Again, the residual is defined as the difference between these images. Similar to I-DIC the model parameters are updated iteratively to obtain a good correlation between the experiment and the simulation, as shown in Fig. 8.1.

## 8.3 Images

In conventional digital image correlation methods contrast in the images is usually provided by some sort of speckle pattern, either naturally present due to the specimen texture or artificially applied with, e.g., spray paint. In the Mechanical Shape Correlation technique the images consist of a projection of the specimen outline. The most elementary projection one could think of is the binary projection, where the background is assigned a contrasting color with respect to the structure itself, e.g., black and white, see Fig. 8.2a. However, since large parts of the image are monochromatic, the residual will only be non-zero in a relatively small area even if correlation is not achieved. This results in a low overall residual and the method will converge too soon. Therefore, the projection chosen for MSC is the signed distance map, i.e., the value for each pixel represents the closest distance to the sample contour, see Fig. 8.2b. With these images the whole field is affected if the unknown parameters in the FE simulation do not characterize the experimentally observed kinematics correctly. Therefore, the residual field is more likely to direct the algorithm to the correct solution.

### *8.3.1 Projection of Finite Element Simulation*

A common type of elements used in FE simulation is shell elements. It deserves some attention as to how to create the projection images from these simulations. Since shell elements are infinitely thin, a projection perpendicular to the structure will not incorporate the real thickness of the sample. Therefore, this thickness should be artificially included in creating the

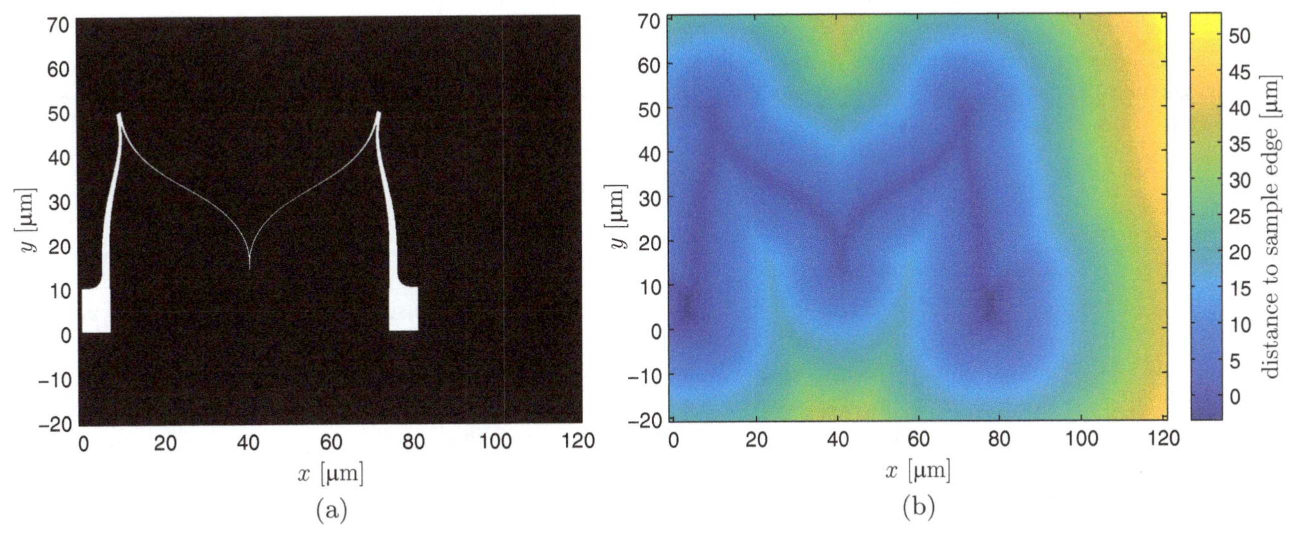


**Fig. 8.2** Projection images used for the Mechanical Shape Correlation method. Two options are shown: on the *left* a binary projection in which the background is *black* and the structure itself is white. On the *right* a signed distance map is shown, where the colors indicate for each pixel the closest distance to the edge of the structure. (**a**) Binary image. (**b**) Signed distance map

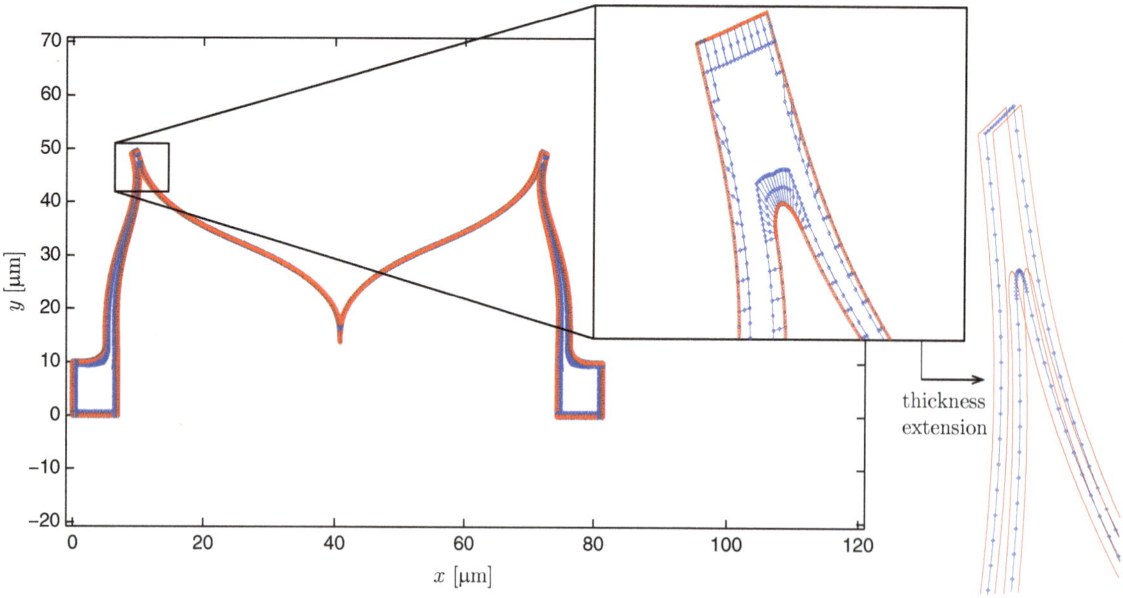


**Fig. 8.3** Determination of the sample contour from an FE simulation. In the case of shell elements, all element corner nodes located on the perimeter of the structure have in common that they are only shared by two or three elements, while nodes located in the center share four elements. Based on this characteristic it is possible to follow the contour of the structure, by starting at some point at the perimeter and repeatedly searching for the next node that satisfies this feature. After the perimeter is determined, the thickness of the specimen is artificially incorporated

signed distance maps. The first step is to determine the perimeter of the model, see Fig. 8.3. This is done by starting at an arbitrary point on the contour, however, it is easiest is to start at a corner. A node in the corner appears in only one elements, hence it is easy to search for such a node. Next all other nodes in this one element are scanned. The characteristic of the nodes located on the model perimeter is that they are shared by only two or three elements (or only one for the corners), while nodes in the center of the structure are shared by four elements. The node in the current element that satisfies this characteristic is the next in line on the contour. Furthermore, it is part of the next element in line, which is then scanned for the next node on the perimeter. This process is repeated until the starting node is attained again. Note that the assumptions made for this method only hold for rectangular elements, for other types the procedure should be adjusted accordingly. Also note that only the corner nodes of the elements are taken into consideration; the nodes at the edge centers, which are present for quadratic elements, are ignored. The next step is to construct the sample edge location more accurately, by defining more points (denser than the number of pixels on this distance). Interpolation using the FE shape functions (e.g., linear or quadratic, depending on the element type) is used to determine the positioning of these points.

Now that the sample contour is accurately defined, the thickness can be incorporated. The rotation information of all nodes, resulting from the FE simulation, is taken into account to determine the locations of the virtual bottom and top contour, see the right part of Fig. 8.3. Now when a projection is made using these edges, and also the side edges connecting the top and bottom contours, the thickness of the sample is incorporated.

## 8.3.2 Masking

As can be seen from the extended structure in Fig. 8.3 many lines are crossing inside the structure, while we are only interested in the total outline since that is the only information that can be obtained from real experimental images. Because the determined edges consist of many points, which can be very close together for sharp edges, it is difficult to determine the convex hull of the shape to locate the outer contour and eliminate the internal curves. Therefore the signed distance map for the pixels located inside the sample geometry can not be accurately determined. Furthermore, even if the signed distance map could be determined accurately, a small deviation in the calculation of the displacement field results in a relatively large effect in the signed distance map and hence the residual inside the structure, since the distances inside the structure are very small. This can lead to numerical instability and poor convergence. An elegant solution is to mask the structure itself and correlate only on the area outside the structure. The impact of masking on the method's performance is tested by means of a virtual experiment in the next section.

## 8.4 Virtual Experiments

Two virtual experiments are executed to test the performance of the MSC method. The first experiment concerns one of the main deformation modes of the considered stretchable electronic interconnects: double bending of beams. In this experiment the effect of masking versus no masking is examined. The second virtual experiment concerns the other main deformation mode: torsion of beams.

### 8.4.1 Double Bending of Beams

The effect of masking is tested by means of a virtual experiment concerning a double bending beam, see Fig. 8.4. In this virtual experiment no real experiment is executed with this structure, but an FE simulation is performed and the projections made from this are used as the 'experimental' images. The model is an elastic model, with only one unknown model parameter: the Young's modulus $E$. The value used for the virtual experiment is 69GPa, or $6.9 \cdot 10^{-2}$N/μm$^2$, since the structure is modeled in micron. This value is the solution which is attempted to be determined using the Mechanical Shape Correlation method. The same model used to create the virtual experiment is also plugged in the MSC algorithm, however a variety of initial guesses for $E$ are given. Also the perturbation factor $P$ is varied. This factor is a parameter in the MSC algorithm, which originates from the derivative of the objective function being needed to minimize the residual. Since it is not possible to do so analytically, the derivative is approximated using a finite difference scheme. The step size for this calculation is taken relative to the value of the considered parameter, i.e., the perturbation factor $P$ is introduced. The quantity of this factor affects the convergence behavior of the method and therefore different values for $P$ are investigated in this research.

Since the real value for the Young's modulus is known, it is possible to determine the error in this parameter. This error is plot against the number of iterations for the situation when masking is applied versus no masking, and for different values

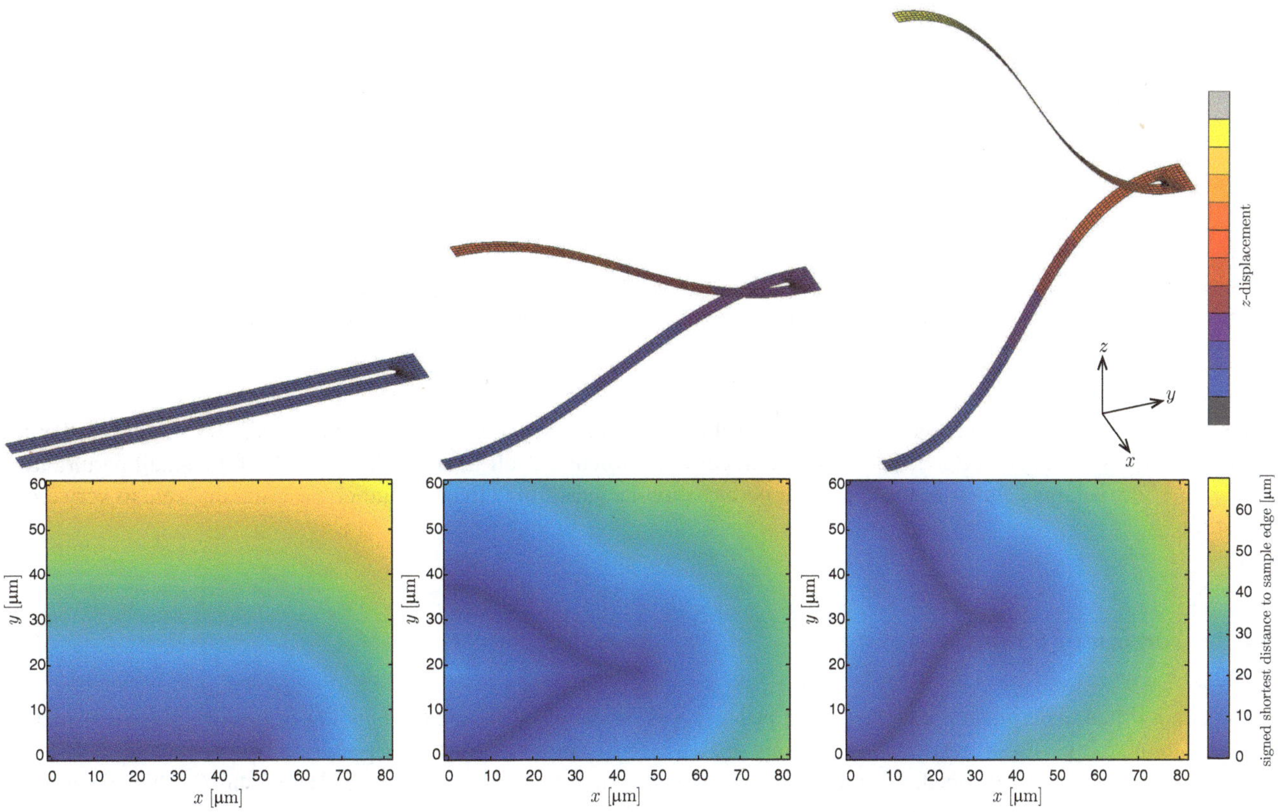


**Fig. 8.4** Virtual experiment with a simple double bending beam. One end of the beam structure is constrained, while the other end is lifted with a prescribed force. The FE model is elastic with a Young's modulus of 69GPa. Not that the $xy$-coordinates in the signed distance maps correspond to the $yz$-coordinates in the FE simulation

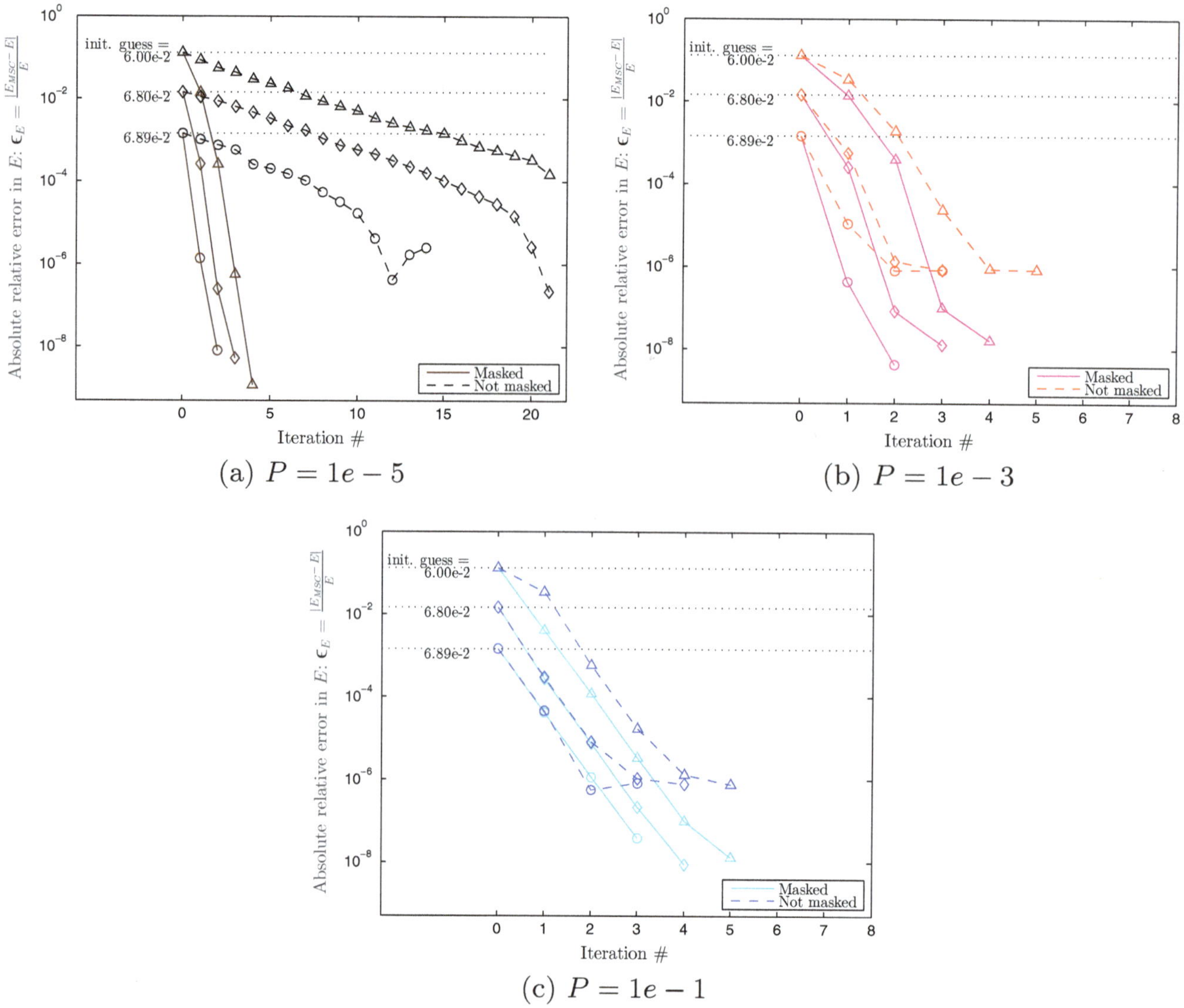


**Fig. 8.5** Comparison of the convergence behavior of the MSC method for images where the structure itself is masked and images for which the structure is not masked, i.e., where the entire image is used. The comparison is executed for different values of the perturbation factor $P$ and for different initial guesses. (a) $P = 1e-5$. (b) $P = 1e-3$. (c) $P = 1e-1$

of the initial guess and perturbation factor $P$. The resulting convergence graphs are shown in Fig. 8.5. It can be seen that for all settings the convergence behavior is improved in case the structure itself is masked. Especially for a small perturbation factor the effect is large. This is because a small perturbation increases the effect of numerical instability due to very small deviations.

## 8.4.2 *Torsion of Beams*

The second virtual experiment is executed for a beam in torsion, which is beside the bending of beams also a deformation mode in stretchable electronic interconnect. A beam of 50 μm × 10 μm is modeled and a prescribed moment is applied on one end, while the other end is clamped. The model is again elastic with one parameter, the Young's modulus $E$, which determines how much the beam rotates upon the prescribed moment. Signed distance projections are made at various increments and these are used as the 'experimental' input images, see Fig. 8.6.

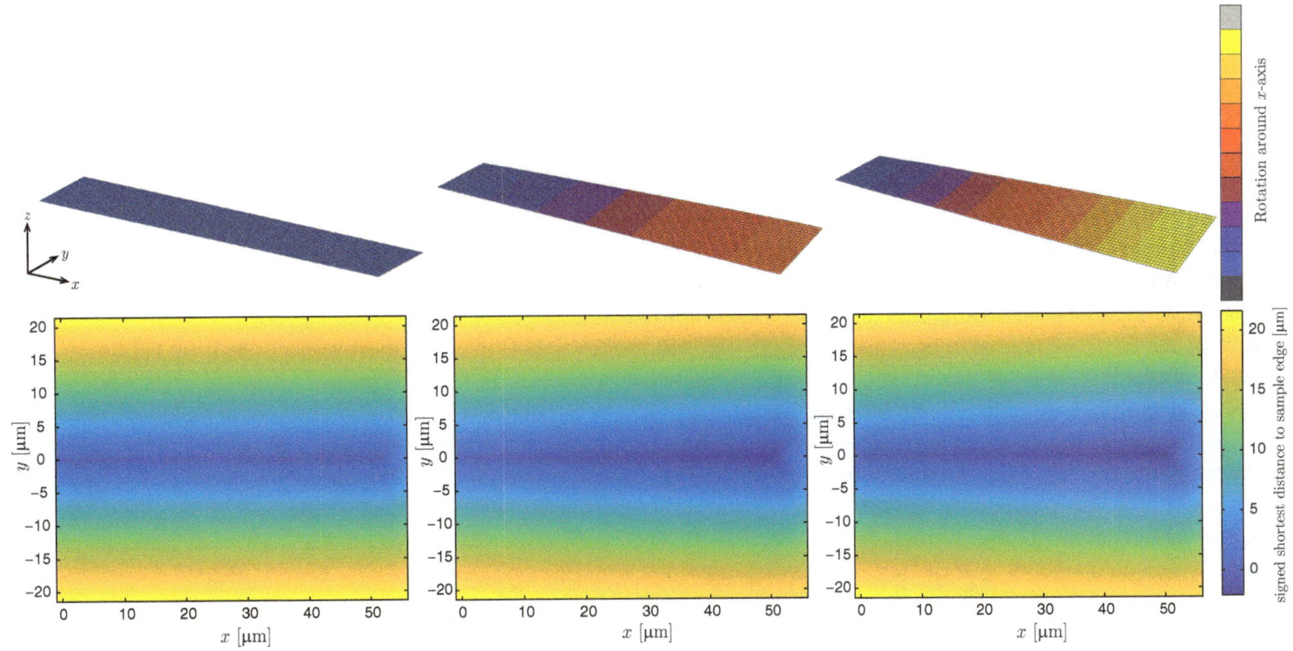


**Fig. 8.6** Signed distance maps for a selection of increments of the virtual experiment concerning a torsion beam. Note that the $y$-coordinate in the signed distance maps coincides with the $z$-coordinate in the FE simulation

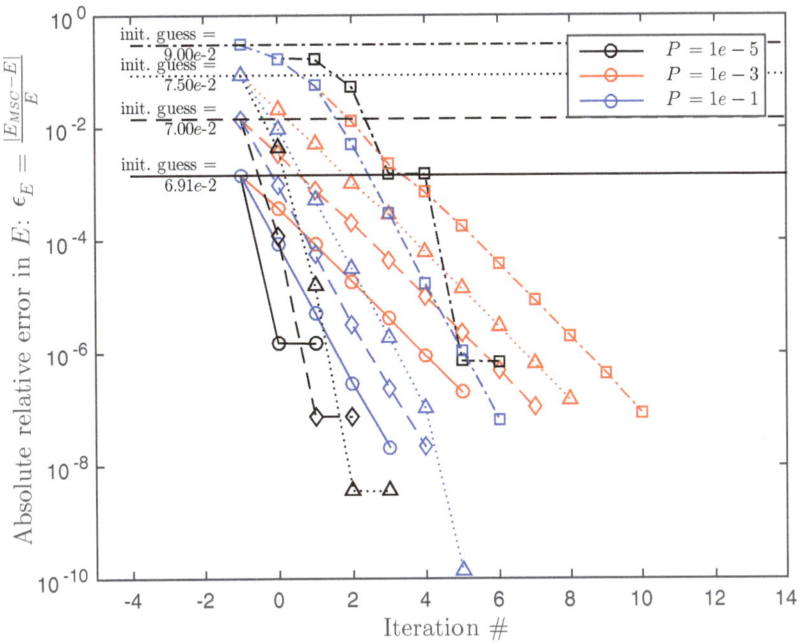


**Fig. 8.7** Convergence plot of the MSC algorithm for the torsion beam experiment, for different values of the perturbation factor $P$ and different initial guess values

The algorithm is run again for different initial guesses and various values of the perturbation factor $P$. The convergence behavior based on the error is plot in Fig. 8.7. It can be seen that convergence with a good accuracy is reached in a reasonable amount of iterations for all tried variations in initial guess and $P$. This gives a good perspective to use the method with other test cases.

## 8.5 Conclusions

A novel method called Mechanical Shape Correlation has been developed. It is an integrated digital image correlation based method, which uses a finite element model for regularization of the displacement field. The goal of such a method is to obtain an adequate set of constitutive model parameters, such that the FE model describes the real, experimentally observed, behavior of the sample correctly. The novelty lies in the use of projections of the sample contour as input images for the algorithm, rather than grayscale speckle patterns. This approach makes it possible to apply the technique to experiments for which existing techniques are difficult to use, because of complex three-dimensional deformations that makes parts of the specimen move out of view or because of complications with pattern application.

The MSC algorithm was explained along with some important changes with respect to the IDIC algorithm. The most important change is the type of images used and attention was paid as to how to create these projections from FE simulations and the important steps taken herein. Also masking of the structure itself in the created projections plays an important role. Two virtual experiments were executed to examine the method's performance. The first virtual test case includes analysis of the difference in results in case the structure is masked in the images or not masked. It was concluded that the convergence behavior improves if masking is applied for all considered parameter variations, including the initial guess and perturbation factor $P$, but that the effect is strongest for small perturbation factors. In both virtual experiments a good accuracy in the objective model parameter, the Young's modulus $E$, was obtained in a reasonably low amount of iterations. These results are promising for the method's performance in other test cases.

## References


1. Neggers, J., Hoefnagels, J.P.M., Geers, M.G.D., Hild, F., Roux, S.: Time-resolved integrated digital image correlation. Int. J. Numer. Methods Eng. **103**(3), 157–182 (2015). doi:10.1002/nme.4882
2. Ruybalid, A.P., Hoefnagels, J.P.M., van der Sluis O., Geers, M.G.D.: Comparison of the identification performance of conventional FEM updated and integrated DIC. Int. J. Numer. Methods Eng. **106**(4), 298–320, (2016). doi:10.1002/nme.5127
3. Kleinendorst, S.M., Hoefnagels, J.P.M., Geers, M.G.D.: Mechanical shape correlation: a novel integrated digital image correlation approach. (2017, In preparation)
4. Han, K., Ciccotti, M., Roux, S.: Measuring nanoscale stress intensity factors with an atomic force microscope. EPL (Europhys. Lett.) **89**(6), 66003 (2010). doi:10.1209/0295-5075/89/66003
5. Bergers, L.I.J.C., Hoefnagels, J.P.M., Geers, M.G.D.: Characterization of time-dependent anelastic microbeam bending mechanics. J. Phys. D. Appl. Phys. **47**, 1–14 (2014). doi:10.1088/0022-3727/47/35/355306
6. Neggers, J., Hoefnagels, J.P.M., Hild, F., Roux, S., Geers, MG.D.: Direct stress-strain measurements from bulged membranes using topography image correlation. Exp. Mech. **54**(5), 717–727 (2014). doi:10.1007/s11340-013-9832-4
7. Kleinendorst, S.M., Hoefnagels, J.P.M., Fleerakkers, R.C., van Maris, M.P.F.H.L., Cattarinuzzi, E., Verhoosel, C.V., Geers, M.G.D.: Adaptive isogeometric digital height correlation: application to stretchbale electronics. Strain **52**(4), 336–354 (2016). doi:10.1111/str.12189
8. Luo, P.F., Chao, Y.J., Sutton, M.A., Peters, W.H.: Accurate measurement of three-dimensional deformations in deformable and rigid bodies using computer vision. Exp. Mech. **33**(2), 123–132 (1993). doi:10.1007/BF02322488
9. Sutton, M.A., Orteu, J.J., Schreier, H.W.: Image Correlation for Shape, Motion and Deformation Measurement: Basis Concepts, Theory and Applications. Springer, Berlin (2009)



**S.M. Kleinendorst** is a PhD-researcher in the Mechanics of Materials group at Eindhoven University of Technology. Her research is about stretchable electronics (SE) and in particular the development of algorithms that couple experimental observations to finite element simulations in order to obtain representative FE models, which can then be used to optimize the SE.

**J.P.M. Hoefnagels** is associate professor in the field of experimental solid mechanics with a focus on thin films and interfaces. The research includes size effects, interface and damage mechanics, flexible/stretchable electronics, and MEMS.

**M.G.D. Geers** is full professor in Mechanics of Materials at the Eindhoven University of Technology, Eindhoven, the Netherlands. His present interests are damage mechanics, micromechanics, multi-scale mechanics, generalized continua, crystal plasticity and metal forming, with a focus on numerical modeling.

# Chapter 9
# On the Boundary Conditions and Optimization Methods in Integrated Digital Image Correlation

**S.M. Kleinendorst, B.J. Verhaegh, J.P.M. Hoefnagels, A. Ruybalid, O. van der Sluis, and M.G.D. Geers**

**Abstract** In integrated digital image correlation (IDIC) methods attention must be paid to the influence of using a correct geometric and material model, but also to make the boundary conditions in the FE simulation match the real experiment. Another issue is the robustness and convergence of the IDIC algorithm itself, especially in cases when (FEM) simulations are slow. These two issues have been explored in this proceeding. The basis of the algorithm is the minimization of the residual. Different approaches for this minimization exist, of which a Gauss-Newton method is used most often. In this paper several other methods are presented as well and their performance is compared in terms of number of FE simulations needed, since this is the most time-consuming step in the iterative procedure. Beside method-specific recommendations, the main finding of this work is that, in practical use of IDIC, it is recommended to start using a very robust, but slow, derivative-free optimization method (e.g. Nelder-Mead) to determine the search direction and increasing the initial guess accuracy, while after some iterations, it is recommended to switch to a faster gradient-based method, e.g. (update-limited) Gauss-Newton.

**Keywords** Integrated digital image correlation • Boundary conditions • Optimization methods • Gauss-Newton • Nelder-mead • Trust-region

## 9.1 Introduction

Interfacial delamination is a key reliability challenge in composites and micro-electronic systems due to (high-density) integration of dissimilar materials. Predictive finite element models are used to minimize delamination failures during design, but require accurate interface models to capture (irreversible) crack initiation and propagation behavior observed in experiments. A generic inverse parameter identification methodology is needed to identify the interface behavior in their as-received state in the micro-electronics component, while it is subjected to realistic loading conditions, such as thermal loading.

Recently, Integrated Global Digital Image Correlation (IDIC) was introduced, which correlates the image patterns by deforming the images using as few as kinematically-admissible 'eigenmodes' as there are material parameters [1] in the interface model [2], thereby greatly enhancing noise insensitivity and robustness [3]. The main challenge lies in that the interface mechanics only generates very subtle changes in the deformation field of the adjacent material layers, therefore, especially high accuracy and robustness in the simulated deformation field is needed, as well as fast convergence because (FEM) simulations including interface mechanics are notoriously slow.

To obtain high displacement accuracy, besides an accurate geometric and material model, precise boundary conditions have often been overlooked. A study into precise boundary conditions for the case of interface mechanics simulations has recently been conducted, see Ref. [4]. Therefore, boundary conditions is not the topic of this proceeding. High robustness and fast convergence can be equally important, hence, these topics are explored here.

The robustness and convergence is determined by the IDIC algorithm. This algorithm is based on the brightness conservation relation, or optical flow relation, which means that material points retain the same brightness upon deformation of the underlying sample. It is with this principle that the displacement field calculated by using the material model is used to back-transform the deformed image to the reference configuration. If the displacement field is calculated correctly, the reference image and the back-transformed image should overlap perfectly (in the absence of noise and algorithm-induced errors such as interpolation) and hence the difference between these two images, denoted by the residual, will decrease to zero. Therefore, the heart of the IDIC algorithm, as well as for other DIC algorithms, is the minimization of this residual.

S.M. Kleinendorst • B.J. Verhaegh • J.P.M. Hoefnagels (✉) • A. Ruybalid • O. van der Sluis • M.G.D. Geers
Department of Mechanical Engineering, Eindhoven University of Technology, Gemini-Zuid 4.122, 5600MB, Eindhoven, The Netherlands
e-mail: j.p.m.hoefnagels@tue.nl



L. Lamberti et al. (eds.), *Advancement of Optical Methods in Experimental Mechanics, Volume 3*,
Conference Proceedings of the Society for Experimental Mechanics Series, DOI 10.1007/978-3-319-63028-1_9

After all, the residual converges to zero if the FE simulation matches the true, experimental, behavior of the specimen and the back-deformed image calculated with this correct displacement field correlates to the undeformed image.

Different approaches exist to solve this minimization problem. Most commonly used in DIC algorithms is a (modified) Gauss-Newton method. This method is relatively fast in the proximity of the solution, but gives robustness issues if the initial guess is farther from the solution. Other methods exist which have their own advantages and disadvantages. A key issue in integrated DIC is the computational cost consumed by the number of FE simulations performed. Every iteration in the optimization method requires at least one FE evaluation. Some methods, however, require the derivative of the objective function, i.e., the residual, towards the degrees of freedom, i.e., the model parameters. This derivative is not analytically available and therefore, often a finite difference scheme is used to approximate the derivative. This calculation requires the number of degrees of freedom amount of FE evaluations, thereby increasing calculation cost and CPU time immensely. Therefore, the different optimization methods are compared based on the number of FE simulations needed to converge.

The lay-out of this paper is as follows. First the virtual experiment used to evaluate the performance of the different optimization methods is introduced in Sect. 9.2. Then, in Sect. 9.3 three different optimization methods are presented and their performance is compared. Finally, conclusions are drawn in Sect. 9.4.

## 9.2 Virtual Test Case

The performance of the different optimization methods is analyzed based on a virtual experiment. This virtual experiment concerns a tensile experiment with a dogbone tensile specimen, see Fig. 9.1, on which a load is prescribed on both edges. In a virtual experiment no real experimental images are used, but a finite element (FE) simulation is executed and the resulting displacement fields are used to artificially deform a reference image, see Fig. 9.2. These images form the input for the integrated DIC method. The objective is to correlate these images in order to find the model parameters, in this case the Young's modulus $E$ and Poisson ratio $\nu$, that were used in the elastic isotropic FE model used to create the images. The parameters used to create the virtual experiment are $E_p = 1.3 \cdot 10^5$ Pa and $\nu_p = 0.28$.

## 9.3 Optimization Methods

Three different iterative optimization methods are tested. The first is a Gauss-Newton method that is often implemented in (integrated) digital image correlation approaches. This is a custom coded method. The other two methods are used from the MATLAB optimization toolbox. The first of these methods is a Trust-Region method with a Newton approach. The last

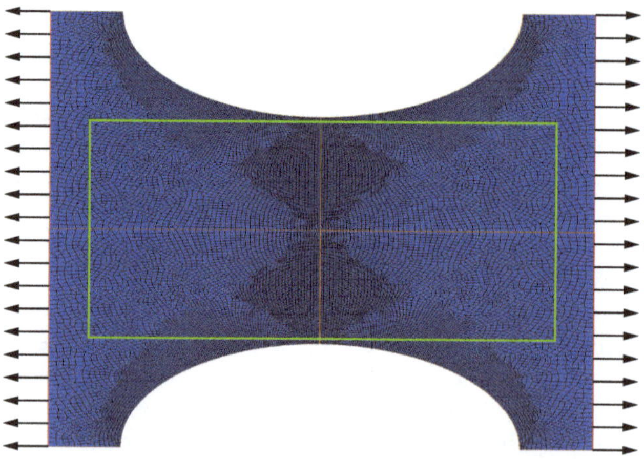

**Fig. 9.1** Finite element simulation for the virtual experiment concerning a dogbone tensile sample. A load is prescribed on the *left* and *right edge*. A quarter of the structure is modeled, the *red lines* indicate symmetry lines. The region of interest is indicated by the *green rectangle*

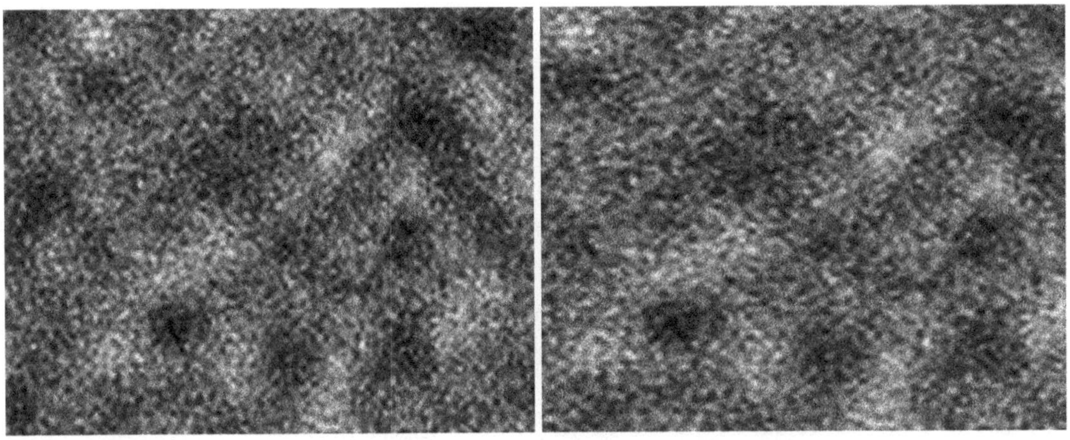


**Fig. 9.2** Reference image $f$ and artificially deformed image $g$. The reference image is a generated speckle pattern with both coarse and fine features, which is suitable for digital image correlation analysis. (**a**) Reference image $f$. (**b**) Deformed image $g$

optimization method is the Nelder-Mead algorithm. The performance of the IDIC algorithm with the different optimization approaches is evaluated by comparing the number iterations needed to converge within a set accuracy. The computational cost is mainly determined by the number of FE evaluations executed and. This number per iterations differs for each method and therefore the total number of FE evaluations for convergence is also compared for the considered methods.

The method are tested for different initial guesses on both parameters $E$ and $\nu$. The initial guesses are given relative to the true material property values used to create the virtual experiment, $E_p$ and $|nu_p$, and are chosen on a logarithmic scale. The set of initial guess values for $E$ ranges from a 10 times lower value to a 10 times higher value. For $\nu$ the range also start with values 10 times lower than the reference value, but it can not increase to a 10 times higher value, since the Poisson's ratio can not exceed 0.5 physically.

Since the iterations are computationally costly, especially for the methods that require derivative information, a maximum of 20 iterations is set. If a initial guess combination for a certain method crosses this limit, but is in the process of converging to the correct solution, this is indicated by a blue color in the graphs. If the method is diverging, an 'x'-symbol is depicted in addition to the blue color.

### *9.3.1 Gauss-Newton*

The Gauss-Newton method is a modification of Newton's method for finding the objective's minimum. Given the set of $m$ image residuals $r = (r_1, .., r_m)$ and the set of $n$ variables $\lambda = \lambda_1, \dots, \lambda_n)$ and the initial guess $\lambda^{(0)}$, the method proceeds by the iterations $s$ described in Equation 9.1 using Jacobian $J_r$ [5].

$$\lambda^{(s+1)} = \lambda^{(s)} - (J_r^T J_r)^{-1} J_r^T r(\lambda^{(s)}). \tag{9.1}$$

This algorithm is often used in digital images correlation methods. Usually $J_r^T J_r$ is denoted as $M$, the optical correlation matrix, and $J_r^T r(\lambda^{(s)})$ as $b$, the right-hand side vector.

To prevent the method from iterating towards non-physical negative values for the model parameters $\lambda$, an update limit it introduced. This limit states that the iterative update in $\lambda$ can not exceed a factor $\alpha < 1$ times the current value of the variable, $\lambda^{(s)}$. Here a factor $\alpha = 0.9$ is used.

The results for a set of initial guesses is shown in Fig. 9.3. Since this method requires the derivative, each iteration requires $(N_{DOF} + 1)$ number of FE evaluations. The method converges for all initial guesses tried, however, for low values of the Young's modulus converges becomes slow. The method is concluded to be relatively robust, but slow if far from the solution.

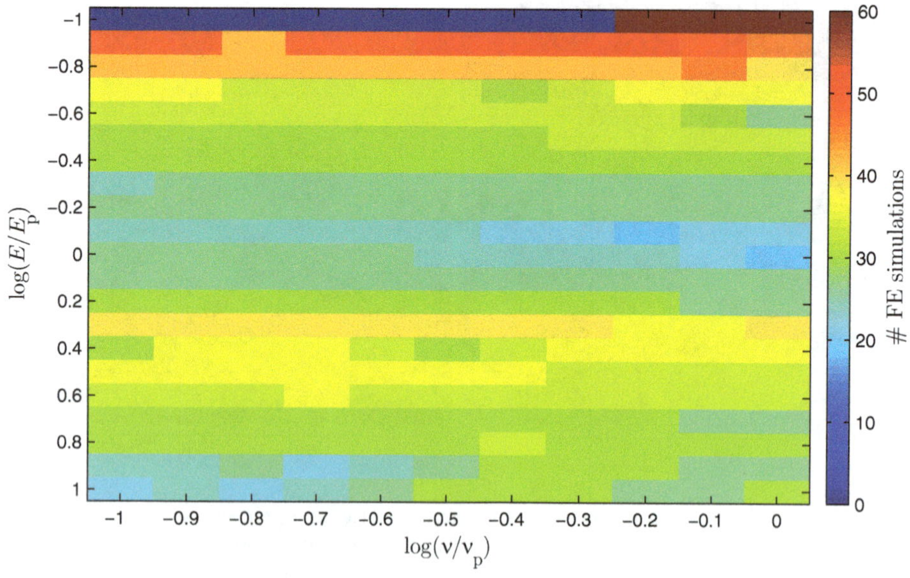


**Fig. 9.3** Number of iterations needed before convergence within a set accuracy limit is reached for the Gauss-Newton method for a range of initial guesses. On the $x$-axis the initial guess in Poisson ratio $\nu$, relative to the reference value $\nu_p$, is plotted. On the $y$-axis the initial guess in Young's modulus $E$, with respect to the reference value $E_p$

### 9.3.2 *Trust-Region*

For the Trust-Region method the MATLAB optimization toolbox is used. The used Optimization Toolbox algorithm is `trust-region-reflective` in the `fmincon` solver. Trust-Region algorithms makes use of trust regions; a defined subdomain $N$ in which the objective function $f$ is approximated with a simpler function $q$ which reflects the behavior of the true objective function. A trial step $s$ is computed by minimizing over this trust region. When the sum of squares is lower for the trial $x + s$, the current point is updated. Otherwise, the current point remains and the trust region $N$ is shrunk. The approximated function $q$ is often the Taylor approximation of $f(x)$. Similar to Gauss-Newton, the Trust-Region algorithm requires an objective gradient and the Hessian Matrix of the posed function, which tends to be slower but more robust to local minima. Using a Trust Region further increases robustness.

Figure 9.4 shows the amount of iterations required to reach the convergence criterium, with 3 (i.e. $N_{DOF} + 1$) FE simulations per iteration due to the calculation of the gradient and Hessian. The Trust-Region method shows less robustness for low initial guesses, but convergence performance is higher than the custom Gauss-Newton method.

### 9.3.3 *Nelder-Mead*

The Nelder-Mead algorithm is a derivative-free method, i.e. no Hessian matrix or gradient is required. The Nelder-Mead method is simplex-based. A simplex $S$ in $\mathbb{R}^n$ is defined as the convex hull of $n + 1$ vertices $x_0, \ldots, x_n \in \mathbb{R}^n$. One iteration of the Nelder-Mead method consists of the following three steps:

1. **Ordering**: Determine the indices $h, s, l$ of the worst, second worst and the best vertex, respectively, in the current working simplex $S$.
2. **Centroid**: Calculate the centroid $c$ of the best side; this is the one opposite the worst vertex $x_h$.
3. **Transformation**: Compute the new working simplex from the current one.

The transformation of the simplex is controlled by four parameters; $\alpha$ for reflection, $\beta$ for contraction, $\gamma$ for expansion and $\delta$ for shrinkage. Figure 9.5 shows resp. (1) reflection, expansion, contraction and (2) shrinking operations. Based on the function values of vertices $x_0, \ldots, x_n$, an appropriate transformation step is chosen. It depends on this choice how many FE evaluations are required per iteration.

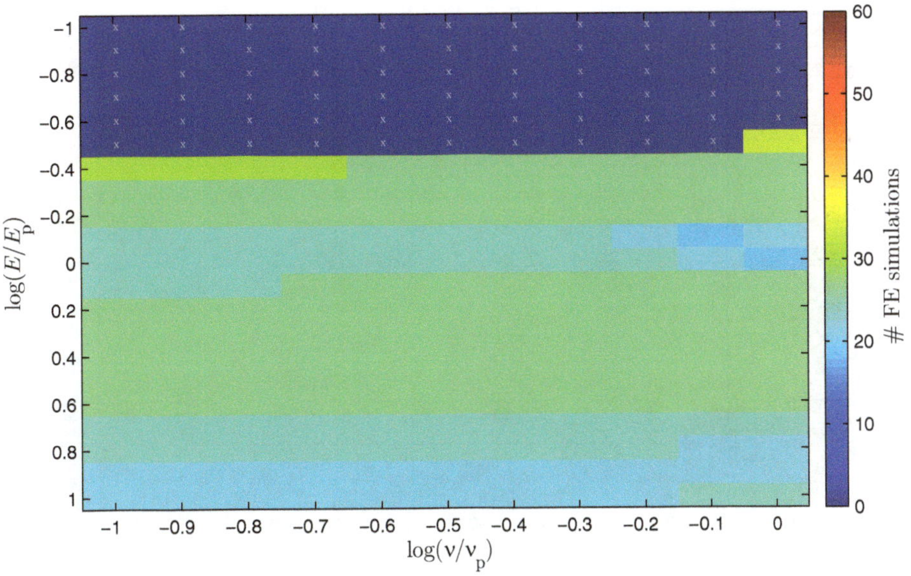


**Fig. 9.4** Number of iterations needed before convergence within a set accuracy limit is reached for the Trust-Region method for a range of initial guesses. On the $x$-axis the initial guess in Poisson ratio $\nu$, relative to the reference value $\nu_p$, is plotted. On the $y$-axis the initial guess in Young's modulus $E$, with respect to the reference value $E_p$

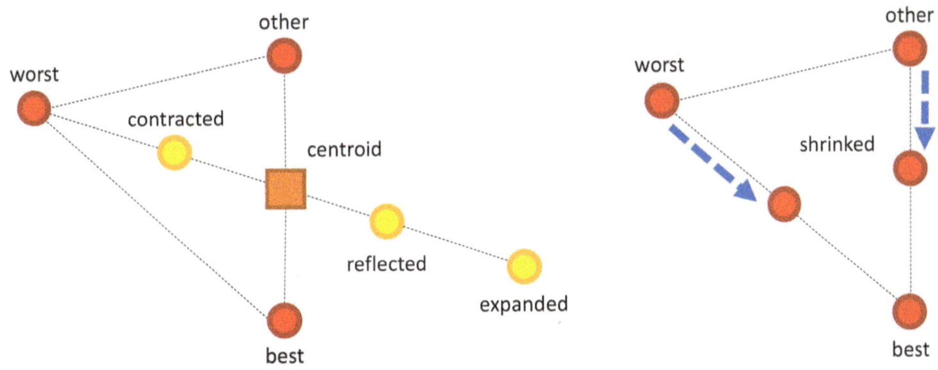


**Fig. 9.5** Operations in the Nelder-Mead algorithm

The following results are for the MATLAB Optimization Toolbox using the Nelder-Mead algorithm. The Nelder-Mead algorithm shows significant convergence towards a minimum for all given initial guesses, which proves robustness, but takes many FE simulations to accurately converge. For all cases in Fig. 9.4, the algorithm was converging, as can be seen in Fig. 9.6.

### *9.3.4 Comparison of the Methods*

To compare the methods more visually, a convergence plot for each method is shown for one specific initial guess combination: $\log(\nu/\nu_p) = -0.1$ and $\log(E/E_p) = -0.3$. Nelder- Mead takes significantly more simulations to converge and cannot rely on a derivative for the search direction, but takes multiple iterations to determine the direction. Gauss-Newton is the fastest method, followed by Trust-Region (Fig. 9.7).

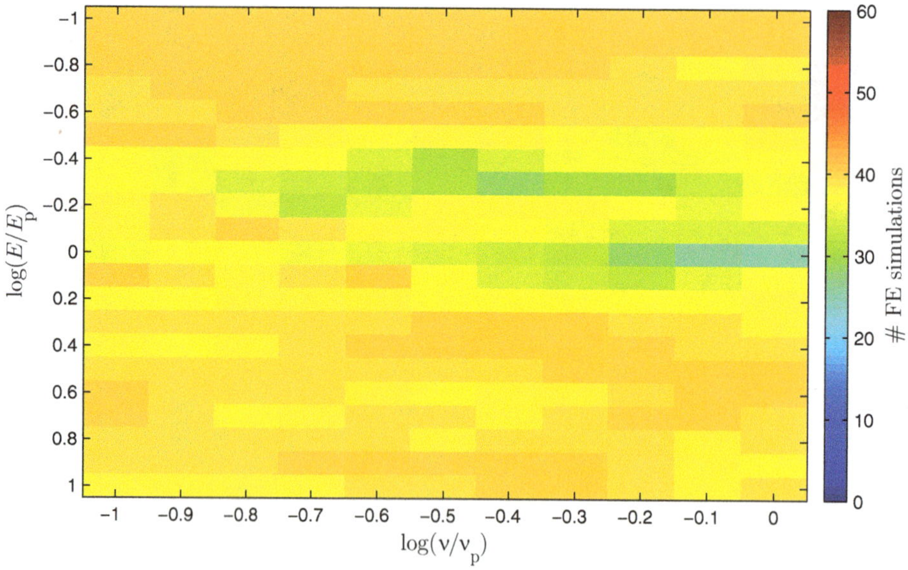


**Fig. 9.6** Number of iterations needed before convergence within a set accuracy limit is reached for the Nelder-Mead method for a range of initial guesses. On the $x$-axis the initial guess in Poisson ratio $\nu$, relative to the reference value $\nu_p$, is plotted. On the $y$-axis the initial guess in Young's modulus $E$, with respect to the reference value $E_p$

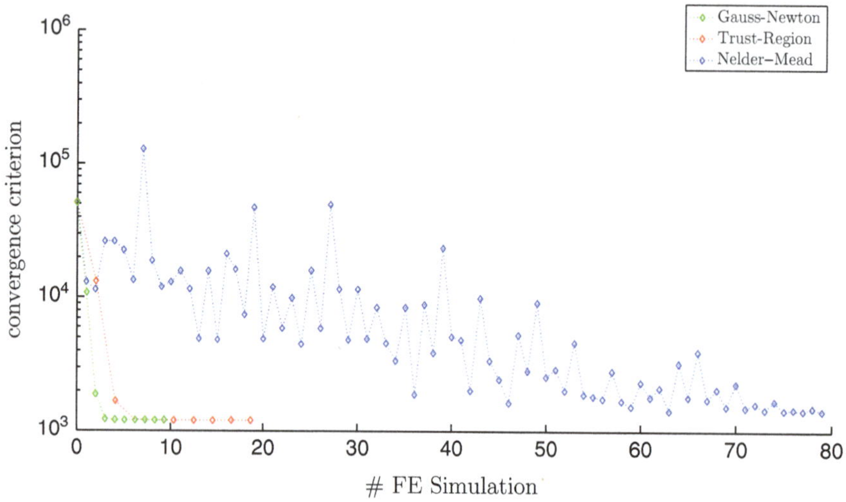


**Fig. 9.7** Convergence plot for all three tested methods

## 9.4 Conclusions

In integrated digital image correlation methods attention must be paid to the influence of using a correct geometric and material model, but also to make the boundary conditions in the FE simulation match the real experiment. Another issue is the robustness and convergence of the IDIC algorithm itself, which has been explored in this proceeding. The basis of the algorithm is the minimization of the residual. Different approaches for this minimization exist, of which a Gauss-Newton method is used most often. In this paper several other methods are presented as well and their performance is compared in terms of number of FE simulations needed, since this is the most time-consuming step in the iterative procedure.

The results presented show varying performance and robustness for different methods. Derivative free methods, like Nelder-Mead, tend to require a great amount of iterations, but it should be noted that this may not be bad for performance, because perturbation of input parameters $\lambda$ is not required and hence less FE evaluations, the most computationally expensive step, are executed per iteration. The custom Gauss-Newton with update limit is a robust and performance-wise a decent solution, though using the MATLAB Optimization Toolbox should be considered because of code reliability and future-proofing, since toolboxes are commercially maintained by Mathworks. Furthermore, implementation of the Toolbox functions enables for using other optimization methods in the future, as new methods are introduced by Mathworks.

Another recommendation is extend the use of an update limit in the custom Gauss-Newton implementation, such that constraints based on physical boundaries are enforced, reducing the solution domain, thereby enhancing the robustness.

In practical use of the IDIC program, when choosing the initial guess is difficult, it can be useful to use a very robust, but slow, derivative-free method (e.g. Nelder-Mead) to determine the search direction and increasing the initial guess accuracy. After some iterations, getting nearer the solution, one can switch to a faster gradient-based method, e.g. Gauss-Newton. Using the implementation of MATLAB Optimization Toolbox, switching optimization methods runtime is possible. The analysis is done using a varying Young's Modulus and Poisson ratio, but IDIC can be used to determine many other constitutive parameters. The different optimization methods may differ in performance and robustness when other mechanical properties are to be identified.

## References

1. Neggers, J., Hoefnagels, J.P.M., Geers, M.G.D., Hild, F., Roux, S.: Time-resolved integrated digital image correlation. Int. J. Numer. Methods Eng. **103**(3), 157–182 (2015). doi:10.1002/nme.4882
2. Blaysat, B., Hoefnagels, J.P.M., Alfano, M., Lubineau, G., Geers, M.G.D.: Interface debonding characterization by image correlation integrated with double cantilever beam kinematics. Int. J. Solids Struct. **55**, 79–91 (2015). doi:10.1016/j.ijsolstr.2014.06.012
3. Ruybalid, A.P., Hoefnagels, J.P.M., van der Sluis, O., Geers, M.G.D.: Comparison of the identification performance of conventional FEM-Updating and Integrated DIC. Int. J. Numer. Methods Eng. **106**(4), 298–320 (2016). doi:10.1002/nme.5127
4. Ruybalid, A.P., Hoefnagels, J.P.M., van derSluis, O., Geers, M.G.D.: Image-based interface characterization with a restricted microscopic field of view. (2017). Submitted for publication
5. Björck, A.: Numerical Methods for Least Squares Problems. Society for Industrial and Applied Mathematics (1996). Retrieved from http://epubs.siam.org/doi/book/10.1137/1.9781611971484. doi:10.1137/1.9781611971484

**Sandra Kleinendorst** PhD-researcher in the Mechanics of Materials group at Eindhoven University of Technology, investigating the optimization of stretchable electronics by developing algorithms that couple micro-mechanical experiments to finite element simulations.

**Bart Verhaegh** Master student at the mechanical engineering department of the Eindhoven University of Technology.

**Johan Hoefnagels** Associate professor at the Eindhoven University of Technology in the field of experimental micro-mechanics with a focus on thin films and interfaces. The research includes size effects, interface and damage mechanics, flexible/stretchable electronics, and MEMS.

**Andre Ruybalid** PhD-researcher in the Mechanics of Materials group at the Eindhoven University of Technology, investigating the use of integrated digital image correlation for characterization of interface behavior in micro-electronic structures.

**Olaf van der Sluis** Associate professor at the Eindhoven University of Technology focusing on numerical modelling of the mechanics of micro-electronics, stretchable electronics, and interface mechanics.

**Marc Geers** Full professor of the group Mechanics of Materials at the Eindhoven University of Technology, with an interest in damage mechanics, micromechanics, multi-scale modelling, generalized continua, crystal plasticity and metal forming.

# Chapter 10
# Extension of the Monogenic Phasor Method to Extract Displacements and Their Derivatives from 3-D Fringe Patterns

**C. A. Sciammarella and L. Lamberti**

**Abstract** Determination of displacement field and its derivatives from fringe patterns entails four steps: (1) information inscription; (2) data recovery; (3) data processing; (4) data analysis. Phase retrieval and processing are very important in fringe analysis. In [1], these steps were discussed for 1D signals, introducing a 2D abstract space as phase evaluation requires a vectorial function. The Hilbert transform allows to obtain signal in-quadrature defining local phase. A 3D abstract space must be generated to handle the analysis of 2D signals and simultaneously extend Hilbert transform to 2D. The theory of a monogenic function introduced in [2] is elaborated here: the 3D vector in a Cartesian complex space is graphically represented by a Poincare sphere. This provides a generalization of the Hilbert transform to a 2D version of the Riesz transform, a modified version of that described in [2]. Theoretical derivations are supported by actual application of theory and corresponding algorithms to 2D fringe patterns and by comparing obtained results with literature.

**Keywords** 2D signals • Displacements and strains of 3D solid bodies • Poincare hyper-sphere • Heart tissue deformation

## 10.1 Introduction

In preceding papers [1, 2], the process of extracting displacement information and its derivatives has been analyzed from the point of view of application of the Image Analysis Science basic framework. Fringe pattern analysis is a sub discipline of the above-mentioned science that is applied to a mathematical model of the kinematics of the continuum. Fringe pattern analysis is also applied in the metrology of bodies. Both fields of deformation analysis and metrology information gathering have many common aspects but the 3D metrology of objects requires the utilization of many additional resources of the Image Analysis Science than the analysis of deformations. This paper will be limited to the analysis of deformations of 3D solids. In [1, 2], a new approach to fringe patterns analysis was introduced. For historic reasons, methods of displacement information determination have been associated with developments of techniques that utilize different approaches to generate and decode signals containing displacement information. Apart from the pathways of different methods, the same basic science is behind the utilized approaches and the same basic rules and restrictions apply to all of them.

## 10.2 Determination of the Displacement Field in 3D

The first step in generating displacement data in 3D is to have a carrier on the volume under observation. A carrier is a known signal that upon deformation of the analyzed body will be modified from a certain known configuration, called the reference configuration or base configuration, to another configuration, the deformed configuration. The displacement information may be extracted from a deterministic signal or from a random signal. The basic methodology is the same, the random signal introduces additional challenges and difficulties but does not modify the process of extracting the displacement

C.A. Sciammarella (✉)
Department of Mechanical, Materials and Aerospace Engineering, Illinois Institute of Technology, 10 SW 32nd St, 60616, Chicago, IL, USA
e-mail: sciammarella@iit.edu

L. Lamberti
Dipartimento Meccanica, Matematica e Management, Politecnico di Bari, Viale Japigia 182, 70126, Bari, Italy



L. Lamberti et al. (eds.), *Advancement of Optical Methods in Experimental Mechanics, Volume 3*,
Conference Proceedings of the Society for Experimental Mechanics Series, DOI 10.1007/978-3-319-63028-1_10

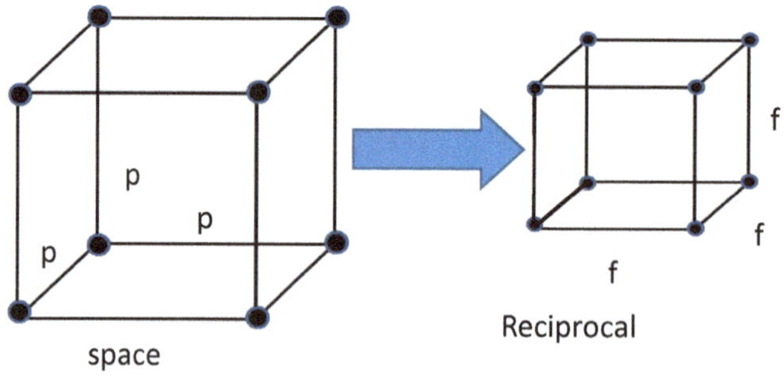


**Fig. 10.1** Physical space tagged with planes separated by the pitch p and reciprocal frequency space with plane distances f

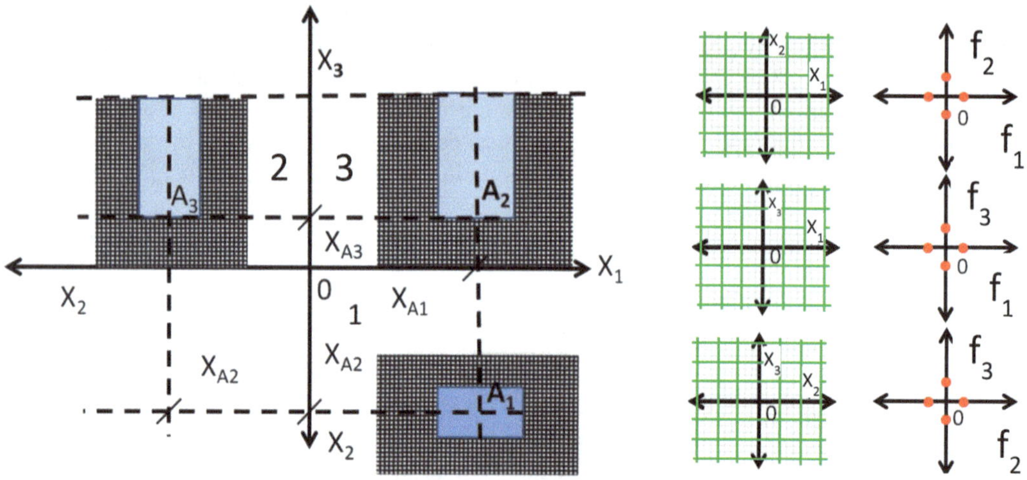

(a) (b)

**Fig. 10.2** A prismatic body tagged with a 3D system of planes; (**a**) Descriptive geometry: 3 quadrants represent sections of the body; (**b**) cross sections of the coordinate planes in space and frequency

information. Let us outline the basic process of extracting 3-D information utilizing deterministic signals. The relationship between $p$ and $f$ is,

$$f = \frac{1}{p} \tag{10.1}$$

Figure 10.1 shows the physical space tagged with planes separated by the pitch p, spatial distance between planes that define the voxels containing the volume information. Corresponding to the voxels in the physical space there are voxels in the reciprocal space or frequency space with plane distances $f$.

Figure 10.2 provides a 3D scheme for spatial and frequency analysis of displacements and the corresponding derivatives. At this point, it is not necessary to specify how the reference system is introduced to the medium under analysis. There are several alternative procedures that can be utilized. The reference system is associated with a Cartesian system of coordinates.

Figure 10.2 illustrates a tagged 3D body utilizing Descriptive Geometry representation. The planes cover the whole volume, Fig. 10.2a. Figure 10.2b shows an enlarged view of the cross sections set of planes corresponding to the voxels that will carry the displacement information and the dots that define in the frequency space the family of orthogonal lines corresponding to these planes.

The coordinate system $X_1$, $X_2$ and $X_3$ is a Lagrangian or material system of reference. The Eulerian system of reference $x_1$, $x_2$ and $x_3$ is selected with planes parallel to the Lagrangian system. To keep the developments very general, large deformations are included, one of the problems that arise is to compare the very different geometries generated in the process of deformation.

Figure 10.3 shows the displacement vector of a voxel and its projections in the coordinate system.

Figure 10.4 illustrates the process of large deformation of a vessel recorded in a time sequence; time t = 0 corresponds to the reference configuration. A parameter must be introduced to connect different stages of the process of deformation,

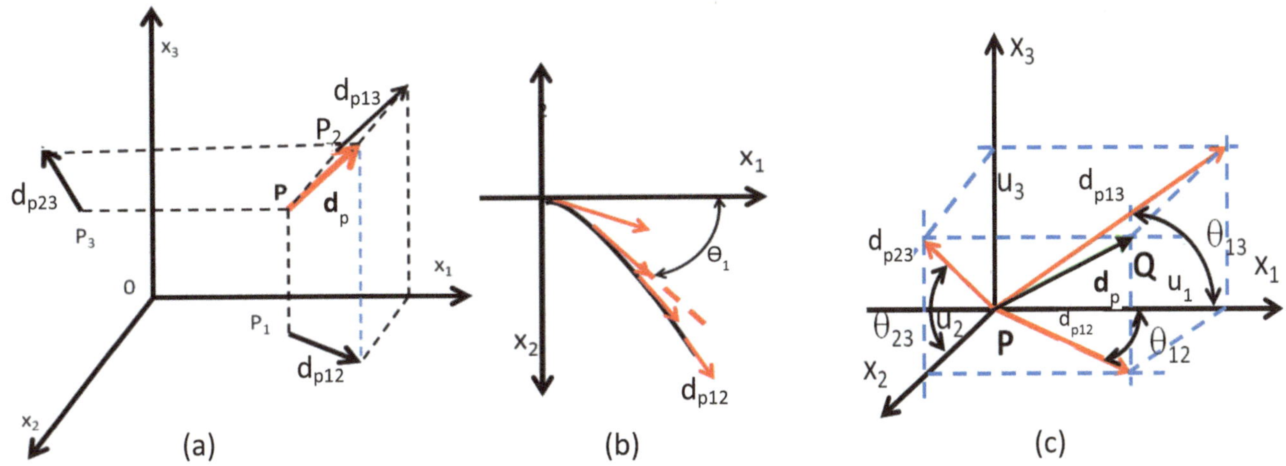


**Fig. 10.3** (**a**) Displacement vector $\mathbf{d_p}$ in the 3-D space; (**b**) projections of the displacement vector in the plane $x_1$-$x_2$, at successive times; (**c**) projections of the vector $d_p$ and corresponding angles

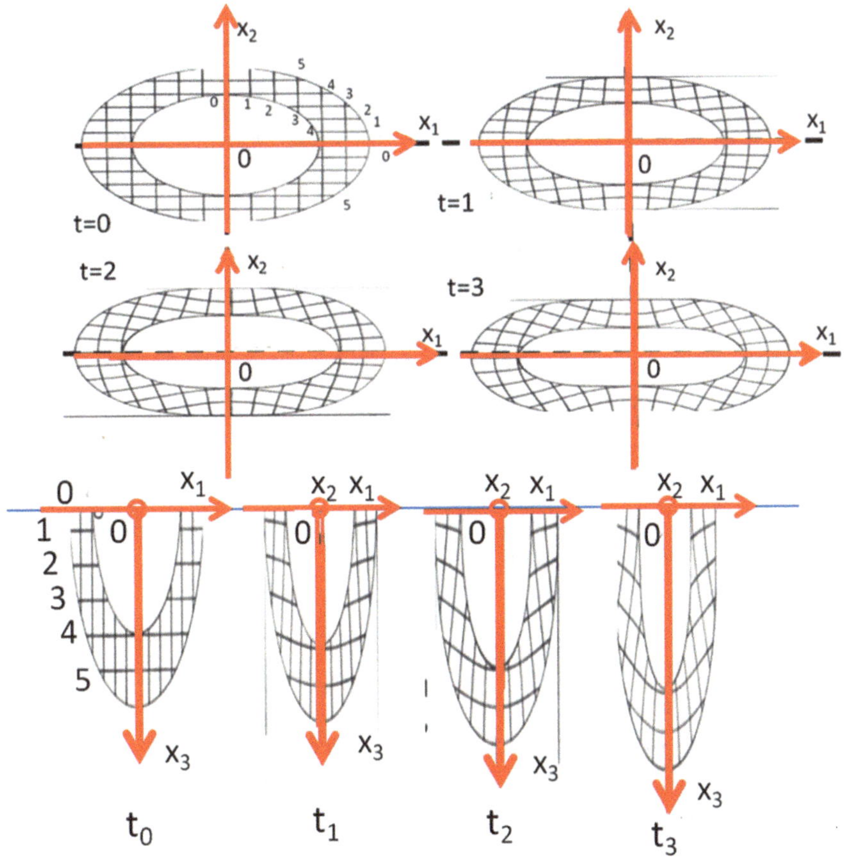


**Fig. 10.4** Large deformations of a vessel taken in a time sequence. Time t = 0 corresponds to the reference configuration

in this example, the time t corresponding to different recordings. Because the planes are fixed to the body's material, the correspondence among volume elements can be obtained due to the parametrization of the planes with orders that are preserved in the deformation course. A matter of notation, we consider real- or complex-valued functions f(**x**) defined on $\Re_n$, where n is an orderly list of integers. Ordinary case letters will represent scalar quantities, bold letters will represent vectorial quantities and we will write f(**x**) or $f(x_1,\ldots,x_n)$, the bold lower case indicating a vector quantity or we will list the low cases variables whichever is more convenient in context. The following notation will be utilized, $f_i(\mathbf{x})$ are functions of **x** and the index "i" will take the values, i = 1,2,3.

Equation (10.2) gives the displacement vector of a point P of the continuum shown in Fig. 10.3, in a Cartesian system of axes,

$$\mathbf{d_p} = u_1 e_1 + u_2 e_2 + u_3 e_3 \tag{10.2}$$

where $e_1$,$e_2$, $e_3$ are the versors of the coordinate system.

The projected vectors in each reference plane are given by,

$$\mathbf{d}_{p12} = u_1\ e_1 + u_2\ e_2 \tag{10.3}$$

$$\mathbf{d}_{p13} = u_1\ e_1 + u_3\ e_3 \tag{10.4}$$

$$\mathbf{d}_{p23} = u_2\ e_2 + u_3\ e_3 \tag{10.5}$$

It is necessary to connect the displacement vector to the signal carrier of displacement information a scalar that represents light intensities in terms of gray levels. This is done through the equation,

$$d_{pij}(x) = \frac{\phi_i(x)}{2\pi} p \tag{10.6}$$

where $\phi_i(\mathbf{x})$ is the local phase, the ratio $\frac{\varphi_i(x)}{2\pi} = n$ is the fringe order. Since we are speaking of local fringe orders, n will be a real number $n \leq 0, 1$. p is the signal pitch. Equation (10.6) is the transformation scale between the intensities and the displacements. Hence, one can write

$$\begin{cases} \left\| d_{p12} \right\| = \frac{\|\phi_{12}\|}{2\pi} p \\ \left\| d_{p13} \right\| = \frac{\|\phi_{13}\|}{2\pi} p \\ \left\| d_{p23} \right\| = \frac{\|\phi_{23}\|}{2\pi} p \end{cases} \tag{10.7}$$

For a generic point P the resultant vector is the sum of three components vectors,

$$\mathbf{d_p}(x, t) = \left[\mathbf{d_{p12}}(x_1, x_2, t) + \mathbf{d_{p23}}(x_2, x_3, t) + \mathbf{d_{p13}}(x_1, x_3, t)\right] \tag{10.8}$$

It can be observed in Fig. 10.3 that the third component $\mathbf{d_{p23}}$ is determined if the other two components are known.

The angle that the projected vector $\mathbf{d_{p12}}$ makes with axis $x_1$ is,

$$\theta_{12} = \operatorname{arctg}\frac{u_2}{u_1} \tag{10.9}$$

Similar equations can be written for $\theta_{13}$ and $\theta_{23}$.

To proceed with the determination of the components of the displacement vector it is necessary to review some of the properties of the Fourier transform, FT, that also apply to the Hilbert transform.

## 10.3 Properties of the Fourier Transform

The recovery of the displacement information is done working with recorded patterns of gray levels (scalar quantities) that result from the modulation of sinusoidal carriers, hence in the following analysis the utilized variables are light intensities I(**x**). In the derivation presented in this section, the multidimensional FT is applied, a generalization of the 1D to multiple spaces that deals with vectorial quantities [3, 4]. While gray levels are scalars, after the FT is applied to gray levels, one gets vector quantities, intensity vectors corresponding to the projection of the displacement vector along a selected coordinate axis. In [2], it was shown that the Fourier transform and the Hilbert transform can be utilized as two possible alternatives to decode displacements from recorded fringe patterns. In this paper, to generalize the procedures derived in [1, 2], we will utilize the FT; similar arguments could be derived applying the generalized Hilbert transform.

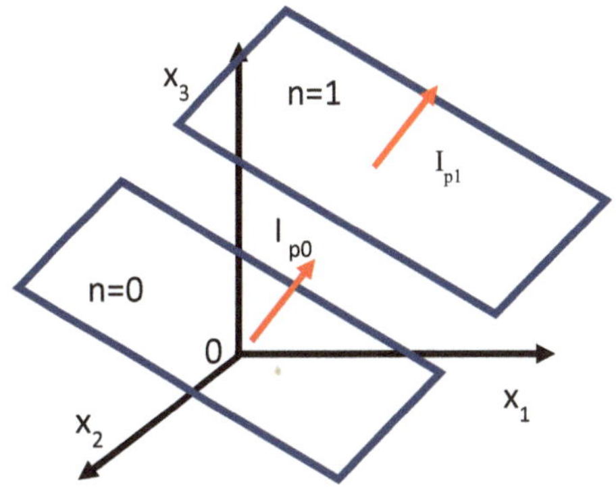


**Fig. 10.5** Planes in the 3D and projected displacement vectors

Let us consider the 3D case, the following equations for the FT applies,

$$\hat{f}(\xi_1, \xi_2, \xi_3) = \mathrm{F}[\mathrm{f}(\mathrm{x}_1, \mathrm{x}_2, \mathrm{x}_3)] = \int_{-\infty}^{\infty}\int_{-\infty}^{\infty}\int_{-\infty}^{\infty} \mathrm{e}^{-2\pi i(x_1\xi_1 + x_2\xi_2 + x_3\xi_3)}\mathrm{f}(\mathrm{x}_1, \mathrm{x}_2, \mathrm{x}_3)\,\mathrm{dx}_1\mathrm{dx}_2\mathrm{dx}_3 \tag{10.10}$$

If we consider the exponent of the exponential function in (10.10), this exponent equals 1 whenever $\mathbf{x} \cdot \boldsymbol{\xi}$ is an integer, that is, when

$$\mathrm{x}_1\xi_1 + \mathrm{x}_2\xi_2 + \mathrm{x}_3\xi_3 = \mathrm{n} \tag{10.11}$$

Equation (10.11) corresponds to the tangent planes to a family of curved surfaces that correspond to the tagging planes shown in Fig. 10.3 and that after deformations have become curved surfaces. These curved surfaces are the equivalent in 3D to the isothetic lines in 2D that locally can be represented by the tangent planes shown in Fig. 10.5. Equation (10.11) represents the orthogonal normal vectors to the 3D isothetic surfaces.

It can be shown that the modulus of the intensity light vector is given by,

$$\|I_p\| = \frac{1}{\sqrt{\xi_1^2 + \xi_2^2 + \xi_3^2}} \tag{10.12}$$

where,

$$\|\xi\| = \sqrt{\xi_1^2 + \xi_2^2 + \xi_3^2} \tag{10.13}$$

$\boldsymbol{\xi}$ is the corresponding frequency vector in the frequency space illustrated in Fig. 10.1.

The exponent of Eq. (10.10) can be written,

$$\mathrm{e}^{-2\pi i(x_1\xi_1 + x_2\xi_2 + x_3\xi_3)} = \mathrm{e}^{-2\pi i x_1\xi_1}\, e^{-2\pi i x_2\xi_2}\, e^{-2\pi i x_3\xi_3} \tag{10.14}$$

Each of the three exponential terms is of the form $\cos\phi + i\sin\phi$;, and the corresponding frequencies are $\xi_i$, ($i = 1,2\ 3$). The first exponential term varies along $x_1$ with a frequency $\xi_1$, likewise the other two terms respectively vary along $x_2$, $x_3$ with frequencies $\xi_2$ and $\xi_3$. This implies three separate transforms along the three axes $x_i$.

The above expression gives the relationship between the vector representing the signal in 3D and its projections in the Cartesian coordinates versors $\mathbf{e}_i$, ($i = 1,2\ 3$). Figure 10.6 represents the 3D vector $\mathbf{I}_p(\mathbf{x})$.

The modulus of this vector is,

$$\|I_p\| = \sqrt{I_1^2 + I_2^2 + I_3^2} = \frac{1}{\sqrt{\xi_1^2 + \xi_2^2 + \xi_3^2}} \tag{10.15}$$

$\mathbf{I}_p$ is the 3D normal vector of the curved surfaces that provide the 3D space equivalent of the 2D isothetic lines. In the 2D case, the projected displacement vector remains in the 2D plane and has components along the $x_1$ and the $x_2$ axes. In the 3D case, the projected displacement vector has components along $x_1$, $x_2$ and $x_3$, Fig. 10.6.

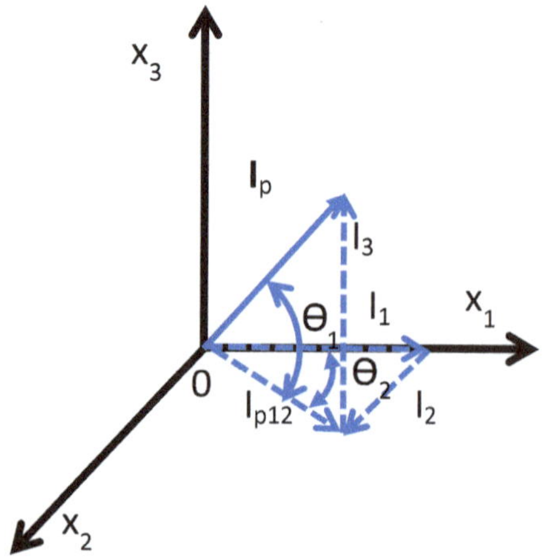


**Fig. 10.6** Resultant projected vector in the 3-D space

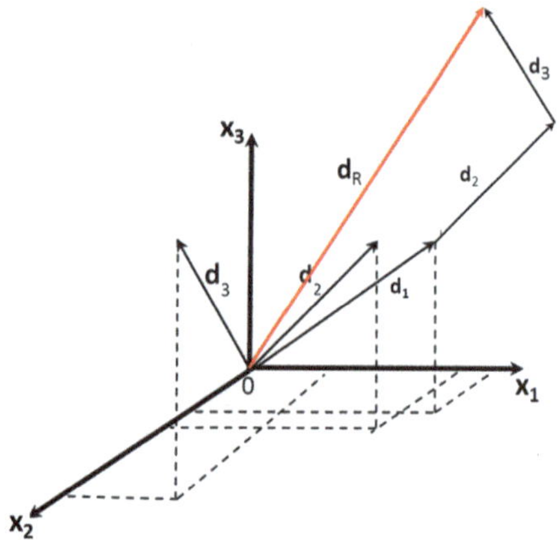


**Fig. 10.7** Vector displacement at a voxel obtained as a sum of the vectors of the family of three tagging planes

From Fig. 10.6 one obtains the relationships between the vector $\boldsymbol{I}_p$ and its projections,

$$\mathrm{I}_1 = \mathbf{I}_p \cos\,\theta_1\,\cos\,\theta_2 \tag{10.16}$$

$$\mathrm{I}_2 = \mathbf{I}_p \cos\,\theta_1\,\sin\,\theta_2 \tag{10.17}$$

$$\mathrm{I}_3 = \mathbf{I}_p \sin\,\theta_1 \tag{10.18}$$

Equations (10.16) to (10.18) give the relationship between the vector $\boldsymbol{I}_p$ and its projections in the Cartesian coordinates of versors $\boldsymbol{e}_i$, (i=1,2 3).

Calling vector $\boldsymbol{I}_{p1}$ one of the three projection vectors ($\boldsymbol{I}_{p1}$, $\boldsymbol{I}_{p2}$ and $\boldsymbol{I}_{p3}$), that correspond to the three families of tagging planes, the resultant vector is given by the vectorial equation,

$$\boldsymbol{I}_{pT}(\boldsymbol{x}) = \boldsymbol{I}_{p1}(\boldsymbol{x}) + \boldsymbol{I}_{p2}(\boldsymbol{x}) + \boldsymbol{I}_{p3}(\boldsymbol{x}) \tag{10.19}$$

For each tagging plane with normal $n_i$, there is a vector $\boldsymbol{I}_{pi}$. The vectorial equation (10.19) provides the vector sum. Then, one has three projection equations for each one of the components of the light intensity vectors $\boldsymbol{I}_{pi}(\boldsymbol{x})$,

$$I_{pT}(\boldsymbol{x}) = \left[I_{p11} + I_{p12} + I_{p13}\right]\boldsymbol{e_1} + \left[I_{p11} + I_{p12} + I_{p13}\right]\boldsymbol{e_2} + \left[I_{p11} + I_{p12} + I_{p13}\right]\boldsymbol{e_3} \tag{10.20}$$

These resultant intensity vectors are converted into projected displacements vectors through the application of the theory of complex signals. Equation (10.20) is graphically represented in Fig. 10.7.

## 10.4 Derivation of the 4D Hyper-Sphere

One can generalize the equivalent of the Poincare sphere in the 3D complex space [2] by resorting to a 4D complex space. The 4D expression for a hyper-sphere of radius $R = I_{sp}$ in the complex space is given by the 4D complex vector,

$$\overset{\Rightarrow}{I_{sp}} = \begin{bmatrix} I_1 \\ I_2 \\ I_3 \\ I_4 \end{bmatrix} \tag{10.21}$$

where the modulus of $\overset{\Rightarrow}{I_{sp}}$ is given by,

$$\left\| \overset{\Rightarrow}{I_{sp}} \right\| = I_{sp} = \sqrt{I_1^2 + I_2^2 + I_3^2 + I_4^2} = R^2 \tag{10.22}$$

The components of the vector are,

$$I_1 = I_{sp}\cos\phi\ \cos\theta_1\ \cos\theta_2 \tag{10.23}$$

$$I_2 = I_{sp}\cos\phi\ \cos\theta_1\ \sin\theta_2 \tag{10.24}$$

$$I_3 = I_{sp}\cos\phi\ \sin\theta_1 \tag{10.25}$$

$$I_4 = I_{sp}\sin\phi \tag{10.26}$$

Figure 10.8 illustrates the relationship between the vector $I_p$ and the vector $I_{sp}$ in the complex plane. Equation (10.15) provides $I_p$.

From Eq. (10.2),

$$\left\| \mathbf{d}_p(x) \right\| = \sqrt{u_1^2(x_1) + u_2^2(x_2) + u_3^2(x_3)} \tag{10.27}$$

Calling $d_p$, the modulus of $\boldsymbol{d}_p$ and the local phase $\phi$, from Eq. (10.6) it follows,

$$\phi = \frac{2\pi d_p}{p} \tag{10.28}$$

From Fig. 10.8,

$$I_{sp} = \frac{I_p}{\cos\phi} \tag{10.29}$$

Then one has obtained all the elements of the complex vector $\overset{\Rightarrow}{\mathbf{I}_{sp}}$.

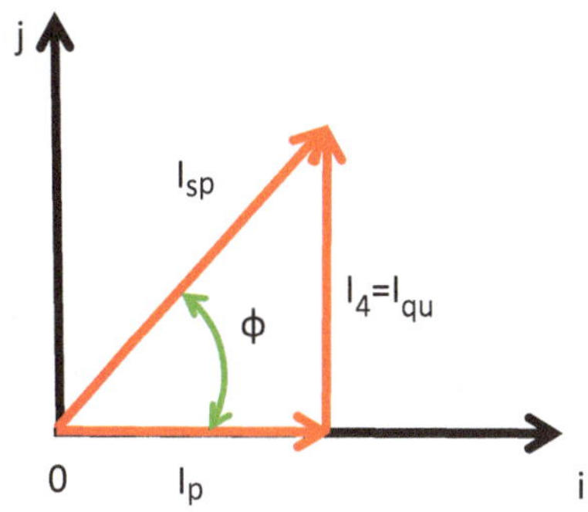


**Fig. 10.8** Determination of $I_4$ in the complex plane

## 10.5 Computation of the Derivatives of the Displacements

In [5], it is mentioned that there are four fundamental tensors that can be utilized in Continuum Mechanics to define different forms of strain tensors and the corresponding definitions are given. The tensors are denoted in the literature as [F], [G], the inverse $[F]^{-1}$ and [J]. It is necessary to address a problem that occurs when it is not feasible to get all the system of tagging planes indicated in Figs. 10.1 and 10.4.

$$[J] = \begin{bmatrix} \frac{\partial u_1}{\partial x_1} & \frac{\partial u_1}{\partial x_2} & \frac{\partial u_1}{\partial x_3} \\ \frac{\partial u_2}{\partial x_1} & \frac{\partial u_2}{\partial x_2} & \frac{\partial u_2}{\partial x_3} \\ \frac{\partial u_3}{\partial x_1} & \frac{\partial u_3}{\partial x_2} & \frac{\partial u_3}{\partial x_3} \end{bmatrix} \tag{10.30}$$

Equation (10.30) gives the derivatives of the components of the displacement vector in Eulerian coordinates with respect to the Eulerian coordinates of the point.

In [5], it is proven that these derivatives are related by the continuity of the medium deformation through relationships, that in the notation of Fig. 10.3c, can be written for the plane $x_1 x_2$,

$$\frac{\partial u_2}{\partial x_1} = \left(1 + \frac{\partial u_1}{\partial x_1}\right) \tan\theta_{12} \tag{10.31}$$

A similar equation can be obtained for the plane $x_1$-$x_3$,

$$\frac{\partial u_3}{\partial x_1} = \left(1 + \frac{\partial u_1}{\partial x_1}\right) \tan\theta_{13} \tag{10.32}$$

For $x_2 x_3$ the equation is,

$$\frac{\partial u_3}{\partial x_2} = \left(1 + \frac{\partial u_2}{\partial x_2}\right) \tan\theta_{23} \tag{10.33}$$

All these derivatives can be computed utilizing the techniques outlined in [2] by differentiation in the frequency plane [6] without unwrapping the phase information.

## 10.6 Determination of Displacements and Strains of the Heart

As an illustration of the proposed approach to 3D displacement and strain determination, images generated by the Magnetic Image Resonance (MRI) method are utilized. MRI images are produced by "illuminating" the observed sample with an intense magnetic field. To this magnetic field, interrogating signals are applied in a temporal sequence that is required to obtain the magnetization information of the different voxels inside the observed object volume. What information is collected? Position of the voxels in the 3D space and the state of magnetization of the voxels. The state of magnetization depends on the chemical composition of the observed voxels thus though MRI it is possible to obtain information on the 3D spatial composition of the observed object. The magnetic resonance law depends on the changing spinning properties of protons in the nuclei of molecules and can only be observed in chemical elements that have different isotopes. An important property of magnetic resonance is the process of tagging the observed volume with planes that modulate periodically the intensity of the resulting image. The tagging planes are attached to the observed volume elements and create the system of reference illustrated in Figs. 10.1 and 10.2. The tagging process is a time-dependent phenomenon and fades away with time, thus limitations arise in the observation process of deformations. One should also realize that the functions of a camera and a sensor recorder in visible light are performed by hardware and software that generates an image output in levels of gray needed by the human observed to understand the content of the output of the MRI system.

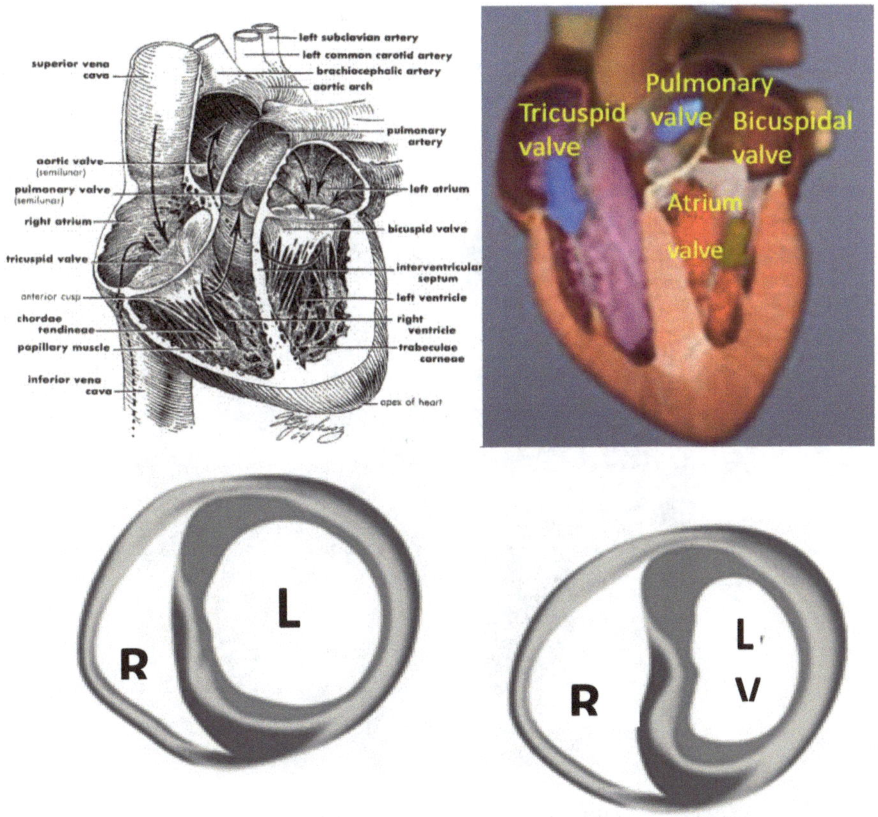


**Fig. 10.9** Heart-long axis cross section (*top*) and short-axes cross sections (*bottom*)

To facilitate the understanding of the deformation analysis of the heart, it is necessary to provide a brief description of the heart anatomy, Fig. 10.9. The heart is a complex pumping system, consisting of two main vessels, right ventricle (RV), left ventricle (LV), a common wall, the septum. At the upper end of the ventricles are located valves controlling the inflow and outflow of blood regulated by electric signals. At each stage of the heart cycle the walls contract or expand to receive the blood inflow and to propel the blood output.

The composition of the heart is complex because it has muscles, veins, arteries, nerves and connecting tissues. For the deformation analysis, the Continuum Mechanics model is utilized and the continuum replaces the actual complex configuration of the heart. If the MRI is recorded from a leaving organism there are problems that must be solved. These problems are consequence of rigid body translations and rotations inherent in a leaving organism. The detailed description of the required correction of the observed images is beyond the scope of this paper. In the images analyzed corrections have been introduced.

Figure 10.10 shows tagged images of the RV and LV of a human heart, it shows the tagged images of the hearts' RV and LV in two subsequent times of the contraction period at mid-systole.

The corrections caused by rigid body motions have already been mentioned. The presence of the tagged lines on the images provided the necessary information to make small corrections for scale and rotation of the images. The reference tagging planes are captured at the instant of time that the heart begins to contract.

For data processing, digital moiré is applied and to get accurate values at boundaries of the ventricles fringe patterns are extended to the full field of view [8]. To complete the data processing [2], additional carrier lines are added to the image. Since the digital moiré is utilized, the fringes in the reference configuration provide the pitch "p" of the gratings utilizing to tag the images. It should be mentioned that orthogonal carriers are not present in the vertical plane, due to the time limitations imposed by the tagging process. As indicated before about Fig. 10.3, only one projection is required. Furthermore, as indicated in the analysis of the derivatives, the missing derivatives can be computed by the corresponding eqs.

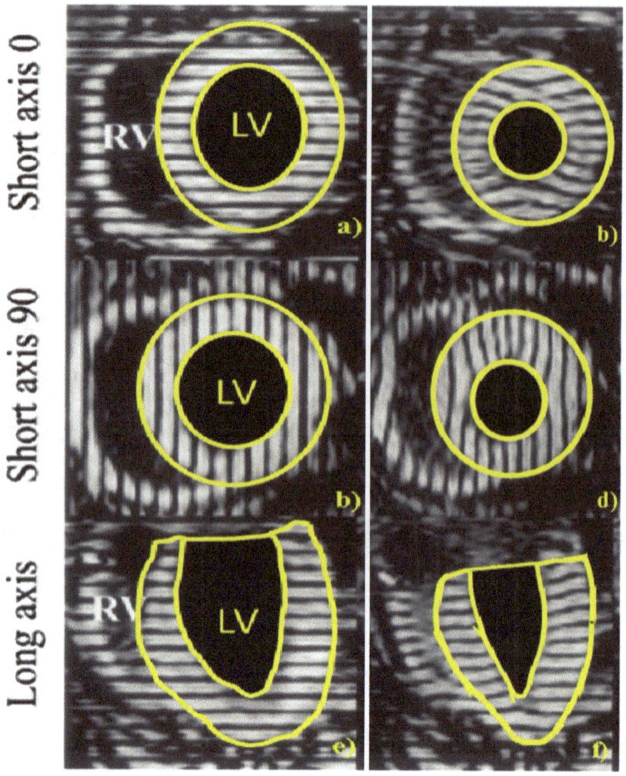


**Fig. 10.10** Short-axis (0° and 90°) and long-axis MRI images of the right ventricle (RV) and left ventricle (LV). Reference configurations correspond to Fig. 10.9a, c, e (Images were taken from Ref. [7])

## 10.7 Processing of the Data Contained in MRI Tagged Images

Figure 10.11 illustrates the process of obtaining the displacements of the heart's left ventricle. In Fig. 10.11, one has the required grid lines patterns $I_u$ that provide horizontal displacements; $I_v$ that corresponds to the vertical displacements and $I_{w1}$ that displays the out-of-plane displacements. Figure 10.10 shows the initial configuration. The components of the displacement $\mathbf{d}_p$ for a point P in the sections A-B and C-D are computed applying the following procedures. The projections of the vector $\mathbf{d}_p$ illustrated in Fig. 10.7, is the sum of $u_1$, $u_2$ and $u_3$. Each one of these components are obtained from the FTs' of the patterns $I_u$, $I_v$ and $I_{w1}$ along axes $x_1$, $x_2$ and $x_3$, respectively. Since the heart is contracting, the arrow direction indicates this fact by going towards P.

The maps of displacements $u_1$, $u_2$ and $u_3$ experienced by the left ventricle are shown in Fig. 10.12a–c. Values indicated by color maps are consistent with the sketch of Fig. 10.11.

## 10.8 Computation of the Strains

From the recorded images and performing the differentiations in the frequency space utilizing the wrapped phase pattern, the tensor of Eq. (10.30) is computed.

The lack of the orthogonal planes in Fig. 10.11e, f implies that some components of the J tensor are unknown, as shown in Eq. (10.34),

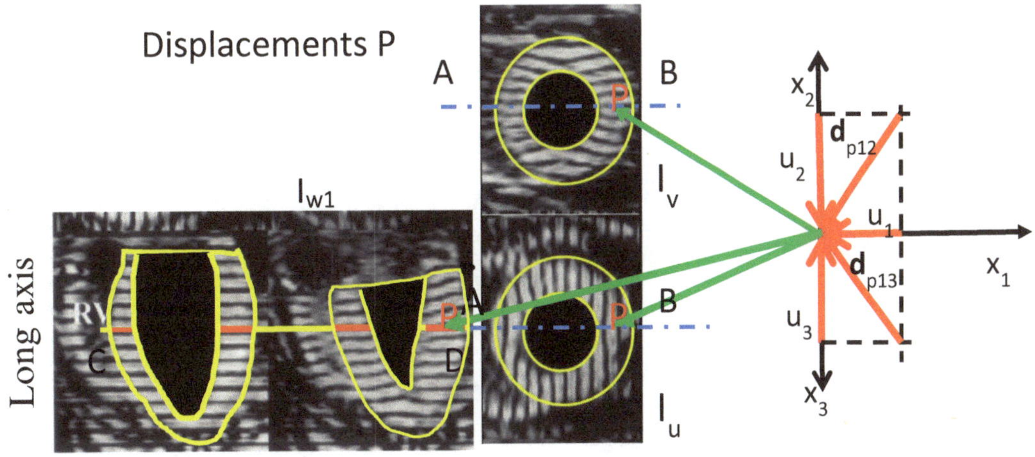


**Fig. 10.11** The figure illustrates the process of retrieving the displacements components at a given point of the left ventricle. The vectors converge to the point P indicating compressive displacements

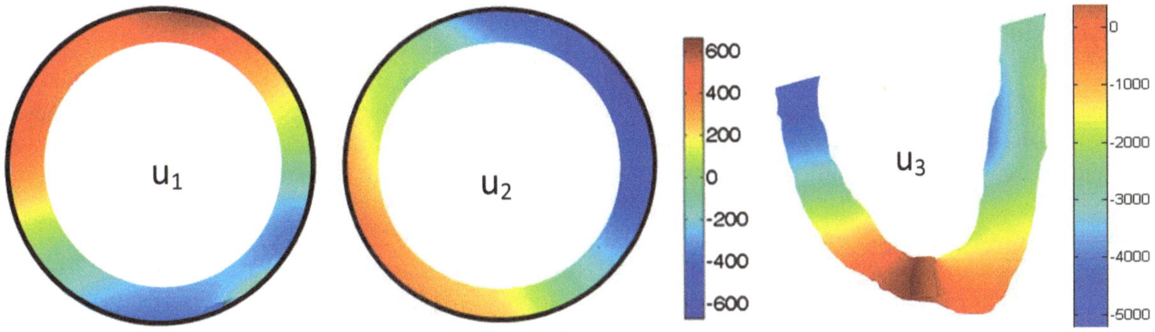


**Fig. 10.12** Displacement components (expressed in μm) of the left ventricle

$$J = \begin{bmatrix} \frac{\partial u_1}{\partial x_1} & \frac{\partial u_1}{\partial x_2} & X \\ \frac{\partial u_2}{\partial x_1} & \frac{\partial u_2}{\partial x_2} & X \\ X & X & \frac{\partial u_3}{\partial x_3} \end{bmatrix} \tag{10.34}$$

These missing derivatives are computed utilizing the relationships among derivatives.

After all the components of Eq. (10.34) are computed, the simplified Almansi strain tensor is obtained. For example,

$$e_{x1}^{E} = \frac{\partial u_1}{\partial x_1} - \frac{1}{2}\left[\left(\frac{\partial u_1}{\partial x_1}\right)^2 + \left(\frac{\partial u_2}{\partial x_1}\right)^2 + \left(\frac{\partial u_3}{\partial x_1}\right)^2\right] \tag{10.35}$$

$$e_{x2}^{E} = \frac{\partial u_2}{\partial x_2} - \frac{1}{2}\left[\left(\frac{\partial u_1}{\partial x_2}\right)^2 + \left(\frac{\partial u_2}{\partial x_2}\right)^2 + \left(\frac{\partial u_3}{\partial x_2}\right)^2\right] \tag{10.36}$$

$$e_{x3}^{E} = \frac{\partial u_3}{\partial x_3} - \frac{1}{2}\left[\left(\frac{\partial u_1}{\partial x_3}\right)^2 + \left(\frac{\partial u_2}{\partial x_3}\right)^2 + \left(\frac{\partial u_3}{\partial x_3}\right)^2\right] \tag{10.37}$$

Figure 10.13 shows the Eulerian components of the simplified Almansi strain tensor along the coordinate axis.

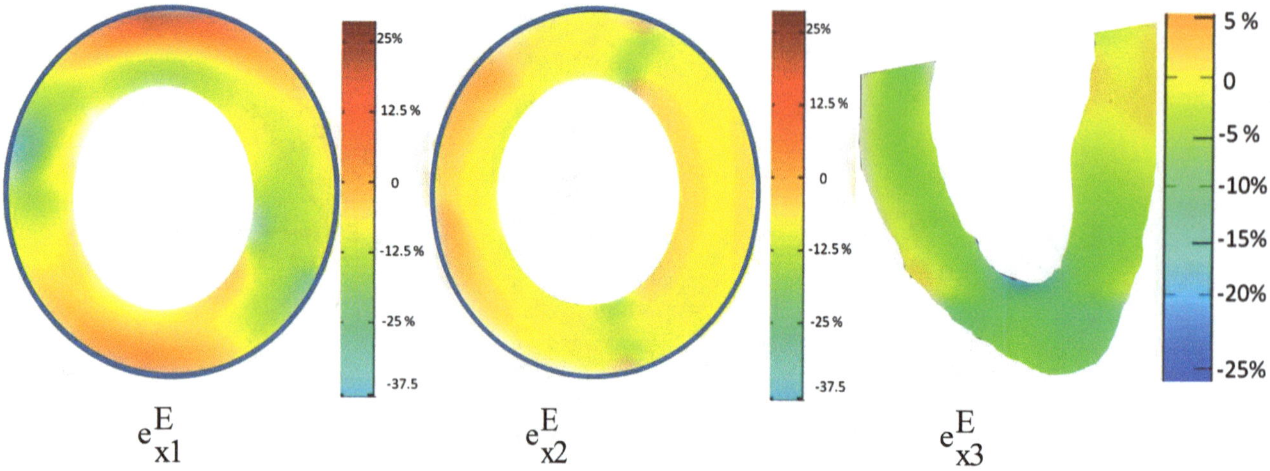


**Fig. 10.13** Eulerian components of the strain tensor along coordinate axes

## 10.9 Discussion and Conclusions

Recently the authors have published papers [1, 2] about fringe pattern displacement information recovery. In these papers were addressed several aspects of fringe pattern processing that required additional analysis and developments to fully achieving the objective of a fast and accurate way of processing signals that contain displacement and deformation information in 2D.

The current paper deals with the generalization of the outlined procedures in 2D to 3D.

In [1], several fundamental aspects of fringe pattern analysis were dealt in the context of 1D signals. It is stressed that a fundamental concept of fringe pattern analysis is the concept of local phase. It is also pointed out that the definition of local phase in 1D requires the introduction of a 2D complex space that is geometrically represented by the unit circle.

In [2], it is shown that the concept of local phase in 2D requires a 3D complex space. Furthermore, it is indicated that the unit circle is replaced by a well-known geometrical surface utilized in Photoelasticity to represent the different states of polarization of light, the Poincare sphere. Thus, the use of the Poincare is extended to patterns associated with displacements and metrological signals, a geometrical illustration of the concept of monogenic phasor. The current paper extends the concept of monogenic function to fringe patterns that correspond to 3D displacements. The addition of one dimension leads to a hyper-4D complex space and to a 4D hyper-sphere. The components of the radius of the 4D sphere are derived. The process of computing the components is illustrated with examples. An application to images produced by the MRI process applied to a human heart left ventricle provides the actual application of the theoretical developments yielding 3D displacements and displacement derivatives. The additional steps of the generalization of the monogenic phasor to the different dimensions are not trivial. Each one of the process of generalization required additional procedures that provided different aspects of the more general problem. For example, the 3D displacements case was handled by introducing a multi-dimensional generalization of the Fourier Transform that deals with vectorial quantities and has very interesting properties that are utilized practically in the process of data retrieval.

The fact that the FT, the Hilbert transform and the Radon transform are connected to each other, opens new avenues for further future developments in the actual study of 3D deformations.

## References

1. Sciammarella, C.A., Lamberti, L.: Mathematical models utilized in the retrieval of displacement information encoded in fringe patterns. Opt. Lasers Eng. **77**, 100–111 (2016)
2. Sciammarella, C.A., Lamberti, L.: Generalization of the Poincare sphere to process 2D displacement signals. Opt. Lasers Eng. **93**, 114–127 (2017)
3. Fourier analysis for vectors. http: www.uio.no/studier/emner/matnat/math/MAT-INF2360/v12/fouriervectors

4. Ciattoni, A., Crosignani, B., Di Portoab, P.: Vectorial free-space optical propagation: a simple approach for generating all-order nonparaxial corrections. Opt. Commun. **177**, 9–13 (2000)
5. Sciammarella, C.A., Lamberti, L.: Basic models supporting experimental mechanics of deformations, geometrical representations, connections among different techniques. Meccanica. **50**, 367–387 (2015)
6. Sciammarella, C.A., Narayanan, R.: The determination of the components of the strain tensor in holographic interferometry. Exp. Mech. **24**, 257–264 (1984)
7. Wyman, B.T., Hunter, W.C., Prinzen, F.W., McVeigh, E.R.: Mapping propagation of mechanical activation in the paced heart with MRI tagging. Am. J. Physiol. Heart Circ. Physiol. **276**(3), H881–H891 (1999)
8. Sciammarella, C.A., Sciammarella, F.: Experimental Solid Mechanics. John Wiley and Sons, Chichester (2012)

**Prof. Cesar Sciammarella** was the Director of the world renowned Experimental Mechanics Laboratory at the Illinois Tech. for more than 30 years. He has received the Theocaris, Hetenyi, Lazan, and Frocht Awards, the William M. Murray Medal and the Honorary Membership of SEM. He received the Fylde Electronics Prize in 2010. He authored more than 200 technical papers.

# Chapter 11
# Deformation Measurement within a Volume of Translucent Yield Stress Material Using Digital Image Correlation

**A. McGhee and P. Ifju**

**Abstract** This paper introduces a method of determining in-plane deformation of a translucent yield stress material (YSM) at any depth using digital image correlation. A 2D plane of uniquely shaped speckles are introduced to a volume of the YSM using a 3D printing technique. A cylindrical object, is dragged through the 2D plane at four different speeds each with four different diameters. The displacements caused by the cylinder were found and analyzed.

**Keywords** 3D printing • Carbomer 940 • Digital image correlation • Soft matter • Yield stress material

## 11.1 Introduction

A robust method of measuring the in-plane displacements within a translucent yield stress material (YSM) has been developed. Although the ability to measure deformation within a yield stress material is nothing new, the techniques employed previously have many limitations which restrict the type of analysis that can be done. Previous experiments have used glass beads suspended in a volume of the YSM which are illuminated with a laser sheet which allow for particle tracking of the glass beads. Since the glass beads are all shaped the same, the displacement of these beads can only be detected if the displacement between an image pair is very low [1]. Large displacements can be measured with this method, but requires many correlations between sequential image pairs to be correlated and summed together; where each correlation introduces some amount of error which would stack up throughout the summation. Our method, however, utilizes uniquely shaped speckles which allow for large scale displacements to be measured directly and with low error. The methods described here can be used with all types of yield stress materials as long as the speckle pattern can be imaged.

The yield stress material used in this experiment is called Carbomer 940 which when introduced to water and a base such as NaOH creates a volume of individually swollen microgel particles with a cross section of about 10 μm. These microgel particles jam together to give the gel material properties of a solid, yet the ability of these particles to shear due to some stress give the gel fluid like properties as well. The amount of stress needed for the particles to shear is called the yield stress, and this yield stress is highly dependent on the rate of the shear event. Rheological analysis show that the Herschel-Bulkley model provide the best fit for the shear rate versus the yield stress [2].

## 11.2 Experimental Method

Digital image correlation (DIC) usually measures the deformation of a surface, but we are interested in the in-plane displacement within a medium. To do this, uniquely shaped speckles made of tinted polyethylene glycol (PEG) are printed into a volume of YSM using a 3D printer with an injection needle attachment (Fig. 11.1). Above the speckle pattern a layer of white paint premixed with the YSM is printed and acts as a background. Cylinders of four different diameters, 0.41, 0.81, 1.27, and 1.82 are moved through the speckle plane at speeds of 0.01, 0.1, 1, and 10 mm/s with the cylinder tip far from the speckle plane.

A. McGhee • P. Ifju (✉)
Department of Mechanical and Aerospace Engineering, University of Florida, Gainesville, FL, 32611, USA
e-mail: ifju@ufl.edu



L. Lamberti et al. (eds.), *Advancement of Optical Methods in Experimental Mechanics, Volume 3*,
Conference Proceedings of the Society for Experimental Mechanics Series, DOI 10.1007/978-3-319-63028-1_11

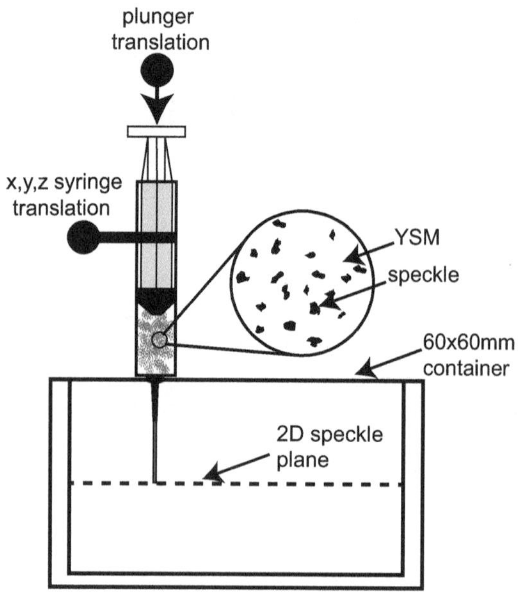


**Fig. 11.1** 3D printing of speckle plane with syringe and 3D printer into container

A 60 × 60 × 40 mm plastic container, filled with YSM, is large enough for boundary effects to be considered negligible. Since the boundary does not interact with the needle, we can make an infinite plane assumption.

A 3D printer is used to insert a 2D plane of polymeric speckles, as well as a white background layer. The same 3D printer used to insert the 2D plane of speckles was used to control the cylinders displacement through the medium. The specimen to be tested was placed on a transparent acrylic surface, and a camera was mounted underneath. A sequence of images were taken as the cylinder is dragged through. These images were then processed using Ncorr (an open source 2D-DIC application for MATLAB) [3]. Using the cylinders location in each deformed image as a reference point, an area around the cylinder was extracted from each deformation measurement. Since the cylinder was at steady state and deforming an infinite plane, all deformation measurements should be the same. By taking many deformation measurements along the path, statistical measurements can be made.

## 11.3 Results

After analysis of each cylinder and speed combination we obtained the u (x-direction) and v (y-direction) displacements as well as a magnitude of displacement plot as can be seen for the 0.81 mm diameter cylinder at 0.1 mm/s in Fig. 11.2. To visualize the effect of velocity and diameter on the displacement away from the cylinder, a plot which shows all speed and diameter combinations is presented in Fig. 11.3. By keeping the scale of the deformation constant in each image, a direct comparison of the effect of velocity and diameter can be seen. By dragging the cylinder back to its initial position and then analyzing the reference image to the total deformation the regions of plastic and elastic deformation can be visualized.

## 11.4 Discussion

The results show high-quality results which measure displacements on the order of 10 μm with a deviation of approximately 2 μm. By increasing camera resolution and frame rate this technique can easily produce higher quality results than those shown. The technique can also be easily modified to measure out-of-plane distortions by introducing a second camera and a 3D correlation processor. Each deformation field in Figs. 11.2 and 11.3.

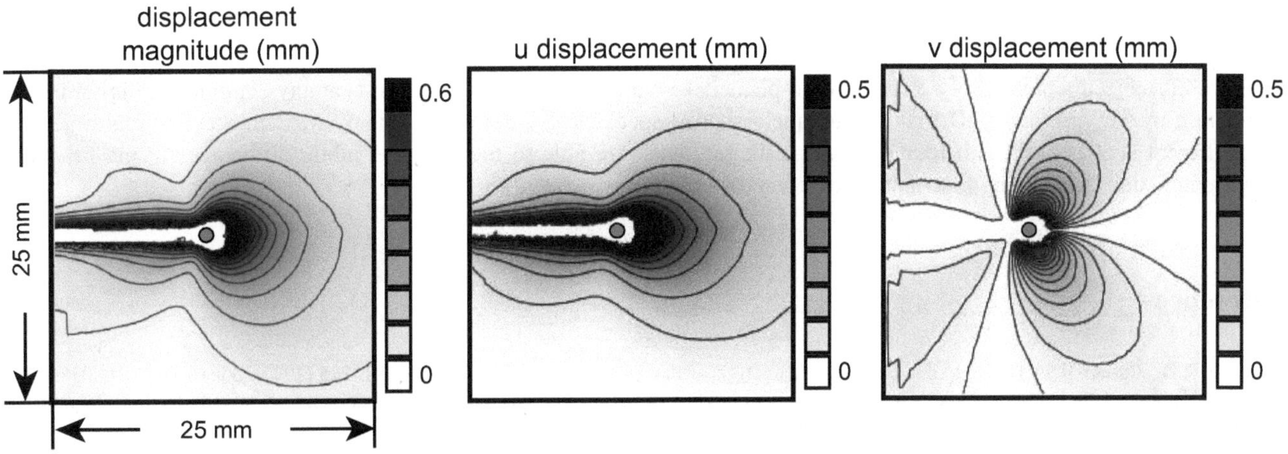


**Fig. 11.2** The resulting deformation fields for the 0.81 mm diameter cylinder at 0.1 mm/s

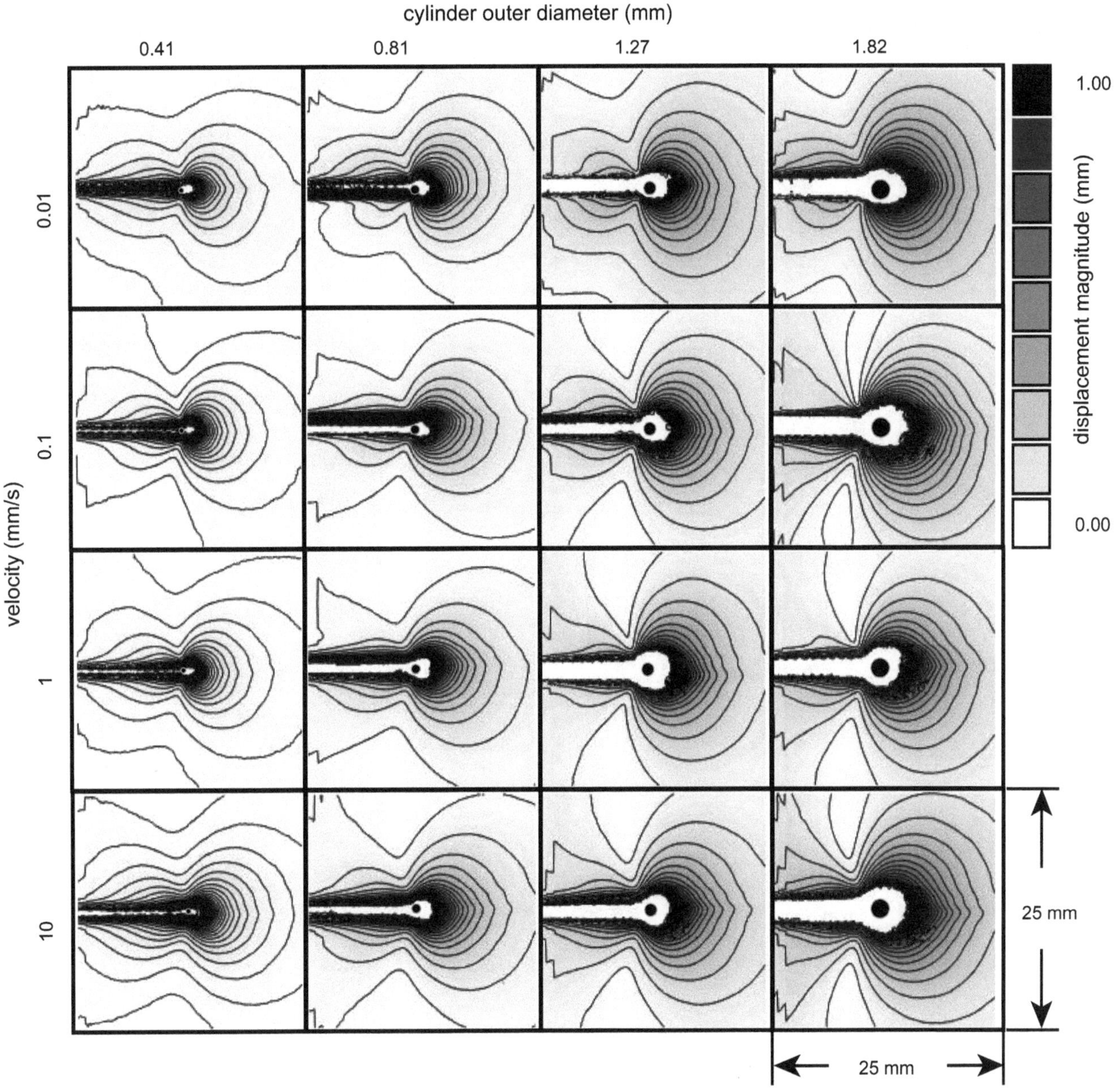


**Fig. 11.3** All cylinder diameters and speed combos plotted with the same scale with contour lines

## 11.5 Conclusion

We have introduced a method of determining in-plane deformation of a translucent YSM) at any depth using digital image correlation by 3D printing a 2D plane of uniquely shaped speckles. By deforming the YSM with a cylindrical object at four different speeds each with four different diameters we were able to measure the subtle differences in the full field displacement due to changing kinematic and geometric conditions.

## References

1. Keane, R.D., Adrian, R.J.: Theory of cross-correlation analysis of PIV images. Appl. Sci. Res. **49**, 191–215 (1992). doi:10.1007/BF00384623
2. Kim, J.Y., Song, J.Y., Lee, E.J., Park, S.K.: Rheological properties and microstructures of Carbopol gel network system. Colloid Polym. Sci. **281**, 614–623 (2003). doi:10.1007/s00396-002-0808-7
3. Blaber, J., Adair, B., Antoniou, A.: Ncorr: open-source 2D digital image correlation Matlab software. Exp. Mech. **55**, 1105–1122 (2015). doi:10.1007/s11340-015-0009-1

**Alex McGhee** is a graduate student in the mechanical engineering department at the University of Florida specializing in experimental soft matter mechanics.

# Chapter 12
# Surface Deformation with Simultaneous Contact Area Measurement for Soft Transparent Media due to Spherical Contact

**A. McGhee, D. Nguyen, and P. Ifju**

**Abstract** We present a method to measure surface deformations between a steel sphere and a flat PDMS surface. A sphere was chosen as the specimen to ensure the resulting deformation measurement can be compared to known theoretical models. A 36 mm diameter steel sphere was pressed into contact against flat, transparent polydimethylsiloxane (PDMS) sheets with a constant load rate controlled by an Instron testing machine. The modulus of the PDMS samples range from 241 kPa to 2.1 MPa. A digital image correlation technique was used to measure the surface deformation of the PDMS with increasing applied load.

**Keywords** Digital image correlation • Frustrated total internal reflectance • Polydimethylsiloxane • Spherical contact • Soft matter

## 12.1 Introduction

A method of investigating contact mechanics between two bodies using digital image correlation (DIC) through transparent media was developed in combination with frustrated total internal reflectance (FTIR). This combination of data enable deformation measurements with the respective contact measurement between the two surfaces. If the object causing the deformation is attached to a universal testing machine (UTM), the displacement and load measurements can be used to produce interesting results such as stress mapped to the contact plot.

The challenges with this method include correction for the index of refraction through the transparent medium as well matching the contact measurement from FTIR to the surface deformation from DIC. This method was created at the University of Florida to measure the contact with corresponding deformation between a hard rough surface and a soft polydimethylsiloxane (PDMS) sheet. To improve and expand upon this method we use a simple spherical surface as the indenter so we can directly compare the results to known deformation models. Furthermore, since the steel ball sticks to the PDMS surface, we measure the pull off deformation and compare it to theoretical models.

## 12.2 Experimental Method

Using an Instron-5969 to control for load rate, a 36 mm diameter steel ball was depressed into the PDMS and then pulled off. The surface deformation and contact area were measured during both the loading phase and pull off phase of the experiment. To measure the surface deformation, a speckle pattern was applied to the contact surface of the PDMS and imaged with stereoscopic cameras. The images were then analyzed using correlated solutions 3D DIC software. The contact area was measured using FTIR. In the FTIR method, a light source illuminates the surface of contact between two objects. Light will reflect internally in a medium until the angle made between the light ray and the surface is greater than some critical angle $\beta$ determined by the relative index of refraction between the boundary. If a body with a higher index of refraction is spaced within a distance equal to the wavelength of light, the light will become frustrated and will become visible on the surface.

A. McGhee • P. Ifju (✉)
Department of Mechanical and Aerospace Engineering, University of Florida, Gainesville, FL, 32611, USA
e-mail: ifju@ufl.edu

D. Nguyen
Department of Civil and Coastal Engineering, University of Florida, Gainesville, FL, 32611, USA



L. Lamberti et al. (eds.), *Advancement of Optical Methods in Experimental Mechanics, Volume 3*,
Conference Proceedings of the Society for Experimental Mechanics Series, DOI 10.1007/978-3-319-63028-1_12

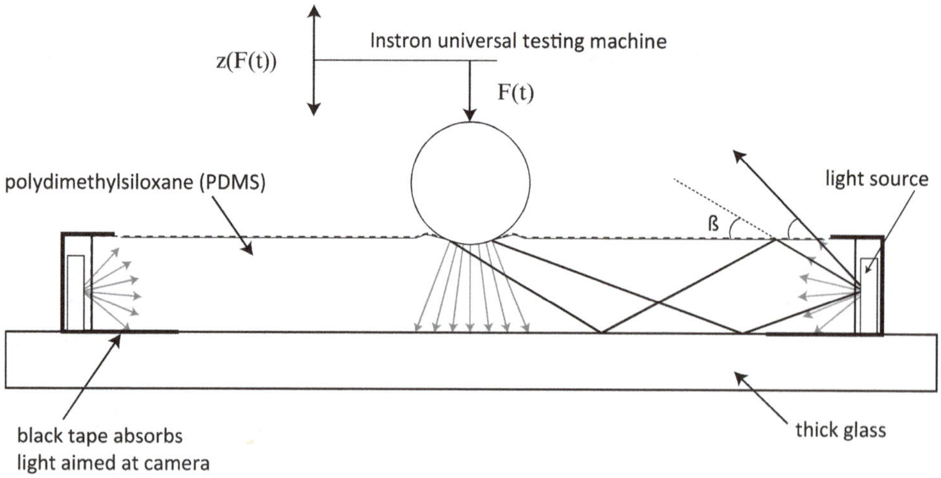


**Fig. 12.1** A cartoon of the experimental setup. The steel sphere is connected to a universal testing machine which applies a load at a constant rate. The rays of light are shown escaping the material at angles greater than $\beta$ and otherwise reflecting internally until the light hits the steel ball and reflects downwards towards the camera

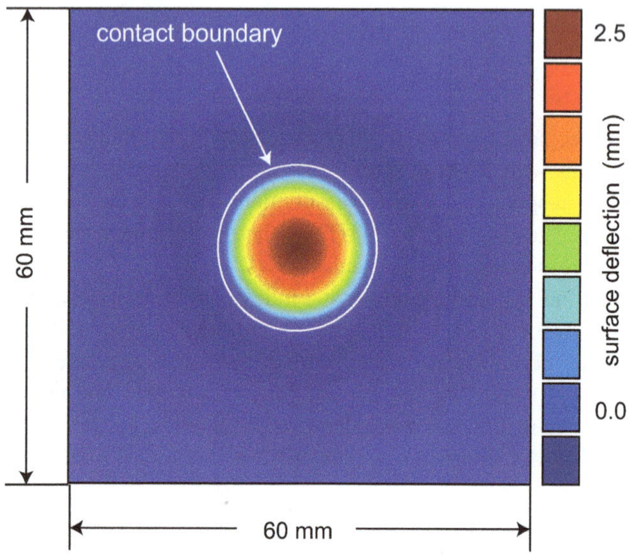


**Fig. 12.2** A surface deflection field from the DIC analysis is presented with the contact boundary from the FTIR analysis shown as a white ring

Using a Nikon D7200 camera mounted directly underneath the specimen, the contact area between the PDMS surface and the steel ball was imaged. The aforementioned experimental setup can be seen in Fig. 12.1. This experimental procedure was done on three different PDMS samples which has a modulus range of 240 kPa to 2.1 MPa.

## 12.3 Results

The results of the analysis shows the outline of the contact boundary overlaid on top of the surface deformation as seen in Fig. 12.2. This deformation field shows the surface deflection in the z direction, with positive values corresponding to indentation and negative values to budging. The raw data of the displacement was larger than the values presented due to the fact that the images were taken through a glass and PDMS surface. To correct for this magnification, a plot of the maximum deflection from the DIC analysis was plotted against the displacement measurements from the universal testing machine and the slope of the given line was found to be 1.35; by dividing the DIC displacement by this slope the surface deflection measurement was corrected.

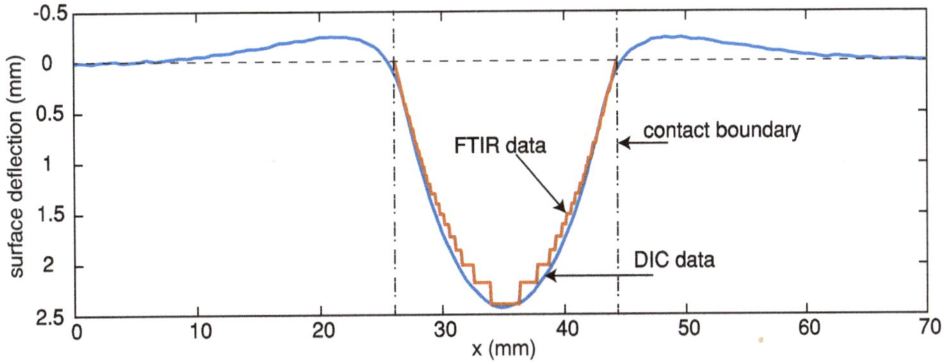


**Fig. 12.3** A plot of the cross section of the surface deflection. The FTIR data was stacked with layers corresponding to the displacement data given by the universal testing machine which allows a 3D surface to be reconstructed from the contact measurement

The FTIR data was used to create a 3D surface by relating the contact area to the corresponding displacement data from the universal testing machine. The result of this analysis is shown in Fig. 12.3 as a cross section of the contact along with the surface deflection measurement of the DIC data. The FTIR data matches up with the surface of the DIC measurement at the edges of previous contact. The contact boundary is found by identifying the edge of the FTIR data.

## 12.4 Discussion

The resulting data found from this method has proven to yield high quality results which can easily be applied to any soft transparent media. Because the shape is so simple, the results can be used to calibrate the displacement data so that other more complex shapes can be tested. The calibration in the radial direction can be found by comparing extensometer displacement, in the radial direction, to the measured deformation. These calibration methods allow us to find the u, v, and w surface displacements which enable stain data to be found.

## 12.5 Conclusion

By indenting a steel ball on a flat PDMS surface with the DIC, FTIR setup, and a universal testing machine to control load rate we are able to obtain a rich set of data which defines the deformation. The analysis of a simple object with known theoretical models for displacement allows for a correction factor to be applied for the u, v, and w-displacements of the surface. This calibration can be used for all other surfaces that are tested on the same surface.

**Alex McGhee** is a graduate student in the mechanical engineering department at the University of Florida specializing in experimental soft matter mechanics.

# Chapter 13
# Towards Measuring Intergranular Force Transmission Using Confocal Microscopy and Digital Volume Correlation

**Kimberley Mac Donald and Guruswami Ravichandran**

**Abstract** We aim to show the feasibility of using confocal microscopy imaging techniques for Digital Volume Correlation (DVC) and analysis of granular mechanics experiments. The first part of this study validates the DVC and confocal microscopy imaging methods for a general problem of uniaxial compression of continuous media. The second part investigates the specific problem of granular mechanics. Intergranular force transmission will be analyzed from 3D image stacks captured using confocal microscopy. DVC methods will be used to analyze the full 3D grain motions and deformations. The intergranular forces will be determined inversely using the Granular Element Method (GEM). Preliminarily results show that confocal microscopy is a useful volumetric imaging method for DVC analyses and shows promise in furthering the study of intergranular force chains and shear bands.

**Keywords** Confocal microscopy • Digital Volume Correlation (DVC) • Granular mechanics • Intergranular forces • Shear bands

## 13.1 Introduction

Intergranular force transmission is of critical importance in analyzing and predicting shear band formation in granular packings. Numerous 2D studies have demonstrated the relationship between shear band slip and the breaking of intergranular force chains. These studies used a variety of analysis methods including photoelasticity and Digital Image Correlation (DIC) [1, 2]. There have also been some studies of 3D granular packings using neutron diffraction and X-ray diffraction methods, which involve considerable effort [3, 4]. These studies demonstrate that further investigation of three-dimensional effects may improve our understanding of the mechanics of granular materials. A potential alternative to neutron and X-ray diffraction methods that allows for capture of volumetric 3D images is confocal microscopy. Commonly used in biological sciences, confocal microscopy uses a series of pinholes to capture images of unique points in 3D space. This imaging method has been used in mechanical studies of cell-matrix interactions [5].

Analysis of 3D studies is often achieved through Digital Volume Correlation (DVC), which is a natural extension of 2D DIC to volumetric space and has been used effectively in both continuum and granular studies [6]. When combined with force inference methods such as the Granular Element Method (GEM), it becomes a particularly powerful tool for analysis of mechanical experiments [7].

## 13.2 Materials and Methods

### *13.2.1 Specimens*

A preliminary study of uniaxial compression was carried out on a hydrogel polymer cylinder. Polyacrylamide specimens were prepared using AMRESCO® ACRYL/BIS™ 19:1 solution with ammonium persulfate (APS, AMRESCO®) and tetramethylethylenediamine (TEMED, Alfa Aesar) as sources of free radicals and as a catalyst respectively. Prior to polymerization, specimens are embedded with 2 μm carboxylate modified FluoSpheres® (580 nm excitation, 605 nm peak

K. Mac Donald (✉) • G. Ravichandran
Division of Engineering and Applied Science, California Institute of Technology, 1200 E. California Blvd, Pasadena, CA, 91125, USA
e-mail: kmacdonald@caltech.edu



L. Lamberti et al. (eds.), *Advancement of Optical Methods in Experimental Mechanics, Volume 3*,
Conference Proceedings of the Society for Experimental Mechanics Series, DOI 10.1007/978-3-319-63028-1_13

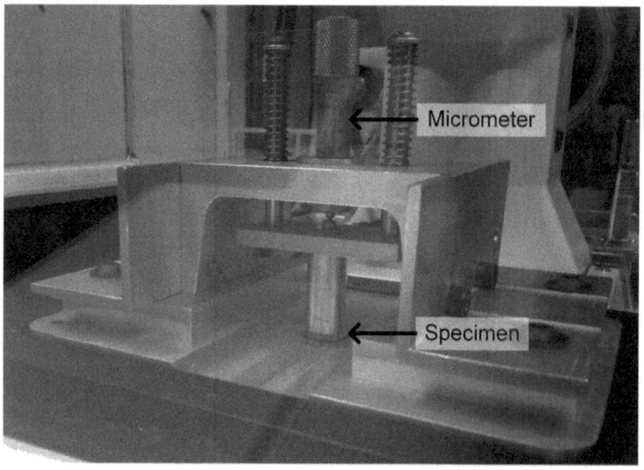


**Fig. 13.1** Compression loading device mounted to microscope stage with polyacrylamide specimen

emission, Molecular Probes™) to produce a DVC speckle pattern. The polyacrylamide is cast into 13 mm diameter molds with a 2 mm height. Initial experiments are carried out on a continuous, homogeneous gel. Specimens with hard poly(methyl methacrylate) (PMMA) inclusions are also studied. A single inclusion is used to validate our methods against an analytical solution [6]. Multiple inclusion interactions are used to further explore the abilities of our methods.

The second part of this study uses model granular packings of fluorescent microspheres. Specimens consist of 180–212 μm fluorescent red polyethylene microspheres (300–550 nm excitation, 605 nm peak emission, Cospheric) in a 13 mm inner diameter polylactic acid (PLA) ring bonded to a 25 mm diameter glass coverslip. These grains are highly spherical and on the scale of fine sand.

### *13.2.2 Experimental Design*

All experiments were carried out on a Zeiss inverted laser scanning confocal microscope (LSM 800) with Zen Blue System software for image stack capture. Images are captured via photomultiplier tubes (PMTs) and laser excitation is achieved using a 561 nm yellow-green diode laser. In these preliminary studies we used a 10×/0.30 M27 EC Plan-Neofluar objective. A uniaxial compression device mounted to a microscope stage insert, shown in Fig. 13.1, is used to apply load via micrometer to all specimens while imaging.

Polyacrylamide specimens are mounted on the compression device using a 25 mm diameter glass coverslip as the base. Unidirectional compression is achieved by applying displacement incrementally with 6–8 min holds for image capture between loading steps. The loading (compression) was achieved in 0.05 mm increments in displacement and the unloading was in 0.01 mm increments. A total of six image stacks were taken for each loading and unloading cycle. The loading cycle stacks each consist of 318 2D images and the unloading cycle stacks consist of 393 2D images taken at 1.0 μm depth increments in both cases. This corresponds to a 317 μm and a 392 μm depth of view respectively.

Granular specimens were mounted in the loading device in a similar manner as the polyacrylamide, but are confined by a PLA ring. To achieve confined compression loading, displacement is applied incrementally at 0.02 mm increments and with 3 min holds for image capture between loading steps. A total of 14 image stacks were taken during loading. Each stack consists of 109 2D images taken at 6.46 μm depth increments, corresponding to a 704 μm depth of view.

## 13.3 Preliminary Results

Analysis of the images of the polyacrylamide specimens show that the achieved speckle pattern is suitable for DIC and DVC analyses. Improvements could be made by adjusting the gain and laser excitation intensity to achieve a more even intensity spread. During analysis of both polyacrylamide and granular specimens, application of point spread function (PSF) deconvolution to correct for optical refractivity effects will allow us to perform more accurate DVC analyses.

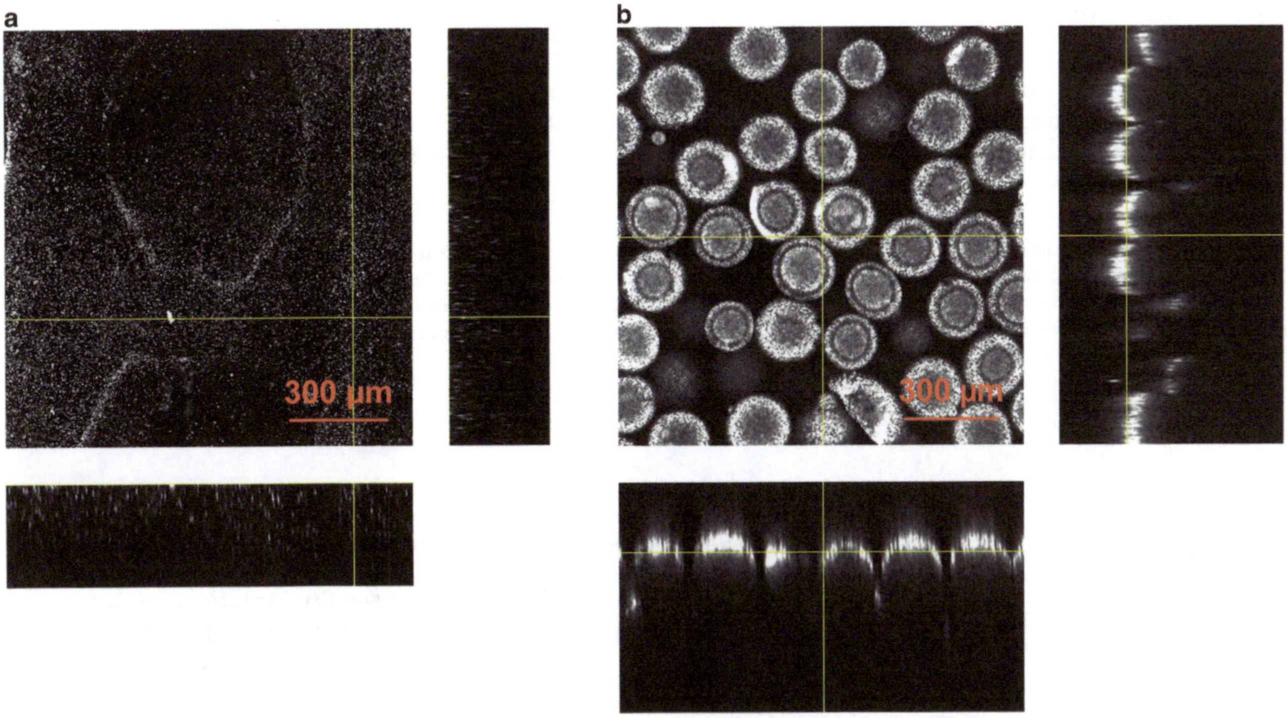


**Fig. 13.2** 2D projections of an image stack of (**a**) polyacrylamide specimen and (**b**) granular specimen

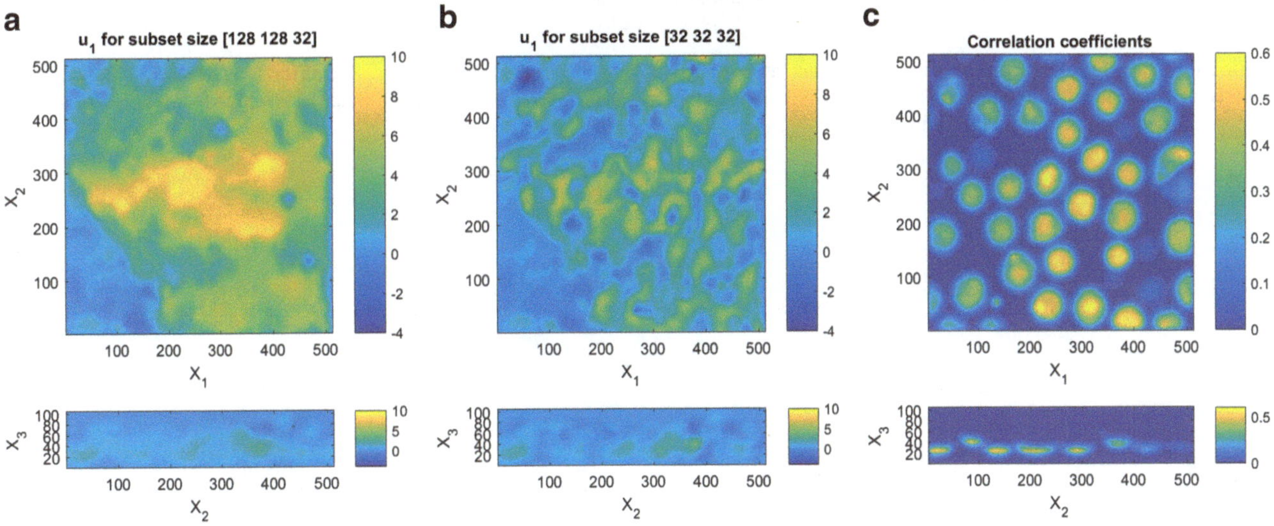


**Fig. 13.3** Displacement component $u_1$ for subset sizes (**a**) [128,128 32] and (**b**) [32 32 32], (**c**) correlation coefficient map

The model granular material shows great promise for image and volume correlation methods using fluorescence confocal microscopy imaging methods. As seen in Fig. 13.2b, individual fluorescent microspheres show strong speckle patterns allowing for both particle tracking and analysis of individual particles. Figure 13.3a, b show that correlation subset size greatly affects the displacement results for DVC analysis due to the large regions of black voxels between individual particles, which are more difficult to correlate. Partitioning of images will allow for greater control over the correlation, as indicated by the correlation coefficient plot in Fig. 13.3c.

Both Figs. 13.2b and 13.3 show one of the major drawbacks to using optical microscopy methods: the depth of view is limited by the optical transparency of the specimen. Particles composed of a more optically transparent medium, lower fluorescent dye intensity, and use of an optically compatible liquid between particles can help to eliminate these effects.

## 13.4 Conclusions

Confocal microscopy offers an alternative method to neutron and X-ray diffraction for capturing volumetric 3D images of granular mechanics experiments. Although there are some drawbacks to the method, it shows promise for analysis of both continuous and granular materials. In particular, we hope to advance experimental methods in the study of intergranular force chains and shear bands.

**Acknowledgements** The project depicted is sponsored by the Department of the Defense, Defense Threat Reduction Agency (HDTRA1-12-0041). The content of the information does not necessarily reflect the position or the policy of the federal government, and no official endorsement should be inferred. This material is based upon work supported by the National Science Foundation Graduate Research Fellowship under Grant No. DGE-1144469. Imaging was performed in the Biological Imaging Facility, with the support of the Caltech Beckman Institute and the Arnold and Mabel Beckman Foundation.

## References

1. Drescher, A., de Josselin de Jong, G.: Photoelastic verification of a mechanical model for the flow of a granular material. J. Mech. Phys. Solids. **20**(5), 337–340 (1972)
2. Hurley, R.C., et al.: Dynamic inter-particle force inference in granular materials: method and application. Exp. Mech. **56**(2), 217–229 (2016)
3. Hall, S.A., et al.: Discrete and continuum analysis of localised deformation in sand using X-ray μCT and volumetric digital image correlation. Géotechnique. **60**, 315–322 (2010)
4. Hall, S.A., et al.: Can intergranular force transmission be identified in sand? Granul. Matter. **13**(3), 251–254 (2011)
5. Franck, C., et al.: Three-dimensional traction force microscopy: a new tool for quantifying cell-matrix interactions. PLoS One. **6**(3), e17833 (2011)
6. Franck, C., et al.: Three-dimensional full-field measurements of large deformations in soft materials using confocal microscopy and digital volume correlation. Exp. Mech. **47**(3), 427–438 (2007)
7. Hurley, R., et al.: Extracting inter-particle forces in opaque granular materials: beyond photoelasticity. J. Mech. Phys. Solids. **63**, 154–166 (2014)

**Kimberley Mac Donald** is currently a PhD student at Caltech studying mechanical engineering. She earned her Bachelor of Science in Architectural Engineering and Master of Science in Civil Engineering from the University of Miami, Florida. Her research focuses on experimental methods in granular and fracture mechanics.

# Chapter 14
# Using Anti-aliasing Camera Filters for DIC: Does It Make a Difference?

**PL. Reu**

**Abstract** Aliased speckle patterns are a known problem for digital image correlation (DIC). By definition, aliased speckles are smaller than the resolution limit of the camera and add "noise" to images via the spatially-aliased frequency content. Aliased speckles occur quite frequently in practical DIC applications, especially when using spray paint to speckle a surface, where control of the speckle size is difficult at best. This paper compares DIC results from aliased speckle patterns imaged with typical machine cameras with and without physical anti-aliasing filters applied to the camera detectors. Additionally, physical anti-aliasing filters are compared with post-processing, digital low-pass filters of aliased images to quantify the influence of the two types of filters on the quality of DIC results. A key result from this work is that the loss of contrast associated with the addition of physical anti-aliasing filters is generally more detrimental to DIC results than the noise resulting from aliased speckles.

**Keywords** Digital Image Correlation (DIC) • Full-field • Optical methods • Uncertainty quantification (UQ) • Aliasing

## 14.1 Introduction

Digital image correlation (DIC) is a full-field displacement and strain measurement technique that uses a digital image of a "speckle" pattern for tracking the underlying motion [1]. The quality of the results relies on the quality of the acquired images, namely the contrast provided by the speckle pattern. This speckle pattern is often applied using a spray paint can and lightly misting the surface. This can create a nice contrasting surface; however, controlling the speckle size is often difficult, and speckles that are too small often result. Additionally, in attempts to increase the spatial resolution of the measurement, speckles are often created that are at the digital sampling limit of two pixels. In both situations, speckles that are smaller than the minimum of two-pixels occur causing the image to be aliased. There is theoretical work on the influence of aliased and undersized speckles on the interpolant and the corresponding bias errors [2]; however, there has been little or no experimental work on this topic. This brief paper looks at an experimental setup that includes three machine vision cameras with two levels of antialiasing filters and an unfiltered camera to investigate the influence of aliased speckles on the DIC results.

## 14.2 Experimental Setup

The experiment consisted of three PointGrey 5-Megapixel cameras with identical Schneider 35-mm lenses. Camera 1 had no antialiasing filters, Camera 2 had a 2-pixel (2Lambda) birefringent blur filter mounted in front of the detector, and Camera 3 had a 4-pixel (4Lambda) blur filter. These filters were designed for the 3.45-μm pixel size of the machine vision camera. To confirm the relative filtering of the antialiasing plates, an Airforce target was used to measure the modulation transfer function (MTF) of the camera and lens system. Figure 14.1 is the MTF of the three systems showing that the resolution of the unfiltered camera is approximately 7.2 line pairs/mm (lp/mm), and the other two are 5.6 and 3.6 lp/mm respectively. The resolution represents the resolving power of the optical system with speckles that are smaller than the resolving limit being filtered out of the image. The cameras were setup to observe a 100-mm field-of-view (FOV) with three different speckle patterns that are observed simultaneously by all three cameras. One pattern was printed with a speckle size of 354 μm or approximately 6-pixels across the speckle to be fully resolved by all three cameras. The aliased pattern was printed with

PL. Reu (✉)
Sandia National Laboratories, 1515 Eubank, Albuquerque, NM, 87123, USA
e-mail: plreu@sandia.gov



L. Lamberti et al. (eds.), *Advancement of Optical Methods in Experimental Mechanics, Volume 3*,
Conference Proceedings of the Society for Experimental Mechanics Series, DOI 10.1007/978-3-319-63028-1_14

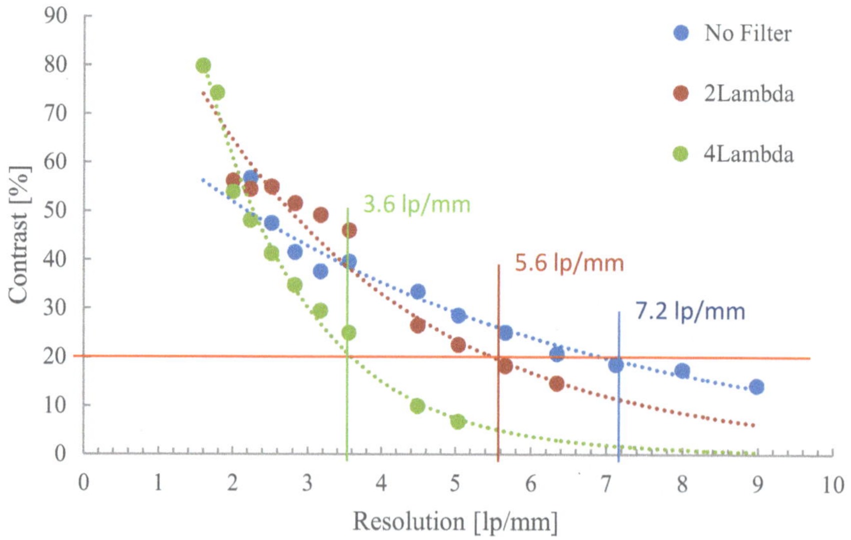


**Fig. 14.1** Modulation transfer function (MTF) of the three camera systems

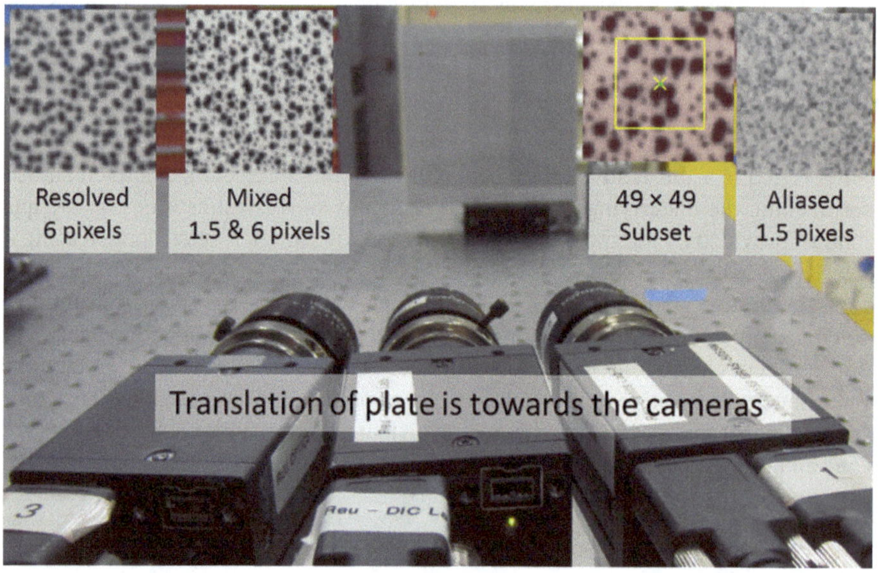


**Fig. 14.2** Experimental setup showing three cameras observing a 100-mm FOV with three different speckle patterns as illustrated in the insert. The DIC subset size is shown overlaid on the mixed speckle pattern a close-up

a 78 μm speckle or 1.5-pixel diameter speckle. The third pattern was a combination of the large and small speckles into a single pattern to represent a situation where there are both aliased and non-aliased speckles, a situation that often occurs when spray painting speckle patterns. The three speckle patterns and the experimental setup are shown in Fig. 14.2. The three speckle patterns were then translated towards the cameras using a linear stage and precision micrometer to create a uniform stretch of the image in the $x$- and $y$-directions. This uniform stretch allows one to easily observe the influence of the interpolation bias error on the results and also provides a uniform displacement and strain field for analysis without the need of expensive precision translation stages.

## 14.3 Results

The images for the three cameras were analyzed using standard subset based commercial DIC codes. The subset size was 49 pixels, the step size was 10 pixels, the strain window was 15 data points yielding a virtual strain gage size of 189 pixels. An affine shape function was used with two different interpolation functions, a bi-linear interpolant and an optimized interpolant. The bi-linear interpolant was chosen because of its poor performance with respect to aliased images. The optimized interpolant was used to show how a more typical (and better) interpolant interacted with the speckle patterns and the three camera images.

Figure 14.3 shows the strain bias errors for the linear interpolant for all three cameras and all three speckle patterns as labeled in the image. The scales are identical showing the relative bias errors between the different patterns and the effect of the anti-aliasing filtering on the results. The tabulated results for the linear interpolant are shown in Table 14.1 and are the peak-to-valley bias error from a line cut in each region of the image. The aliased speckle pattern is clearly the worst for linear interpolants for all three cameras. However, the anti-aliasing filter does not improve the results as may be expected. This is because the loss of contrast from the filtering of the under-resolved speckles has a more dramatic influence on the results than the deleterious influence of the aliased speckle information.

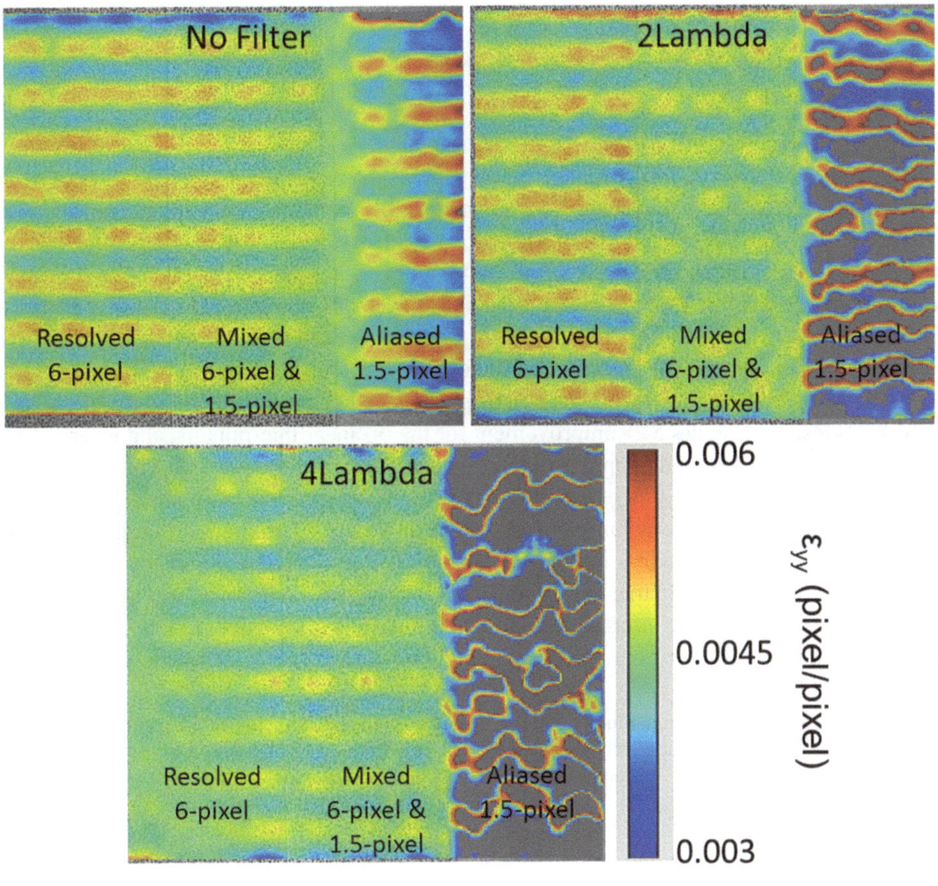


**Fig. 14.3** Strain results ($\varepsilon_{yy}$) for all three cameras and all three speckle patterns with a bi-linear interpolant. Scales are identical showing the relative magnitude of the interpolation bias error in the strain field

**Table 14.1** Strain bias errors (in microstrain) caused by the aliased images using a bi-linear interpolant. Peak-to-Valley strain error is reported

| Filter | Aliased Speckle | Resolved Speckle | Mixed Speckle |
|---|---|---|---|
| None | 4520 | 1309 | 1388 |
| 2Lambda | 5393 | 1182 | 808 |
| 4Lambda | 7437 | 590 | 589 |

Table 14.2 Strain variance error using the optimized interpolant

| Filter | Aliased Speckle | Resolved Speckle | Mixed Speckle |
|---|---|---|---|
| None | 100 | 49 | 33 |
| 2Lambda | 333 | 56 | 85 |
| 4Lambda | 899 | 41 | 43 |

For the fully resolved speckles, the anti-aliasing filter improves the results for the bi-linear interpolant, and has no influence with the optimized interpolant. This is most likely because there is no loss of contrast, because the speckles are fully resolved; however, the edges of the speckles, which in the printed image have hard edges, are filtered and softened. This improves the ability of the interpolant to be able to fit the contrast gradients. Table 14.2 presents the tabulated results of the strain variance in the three speckle regions for the three cameras. There were no noticeable bias errors for the optimized interpolant.

For the mixed speckle patterns, with both aliased and fully-resolved speckles, anti-aliasing does seem to improve the results for the linear interpolants. This is because removing the aliased information helps, while there is less loss of contrast due to the presence of the fully-resolved speckles. For the optimized interpolant, there is no improvement from the anti-aliasing filters.

## 14.4 Conclusions

It has been known that aliased speckles, and speckle size in general, has an influence in the quality of DIC results. This short study confirms those results. Aliased speckles, for either an optimized or bi-linear interpolant (and presumably all interpolants) negatively impact the DIC results. For bi-linear interpolants, where the bias errors are larger, the results are clearly seen in the greater peak-to-valley errors for the aliased patterns regardless of whether there is an antialiasing filter. For the optimized interpolant, the strain noise is worse for aliased speckles, again regardless of whether there is filtering. However, for the optimized interpolant, adding the antialiasing filter actually made the results much worse than having no filter when aliased speckles were present. This was true of the bi-linear as well but to a much smaller extent. This is because the loss of contrast in the image due to filtering out the small speckles is worse than the influence of the speckle aliasing.

For fully-resolved speckles, filtering has a 2× improvement for the bi-linear interpolants, and a very modest improvement on the optimized interpolants. This was again approximately true for the mixed speckle pattern.

These results were somewhat surprising as it was theorized that removing aliased content would greatly improve the results. This is only true when a "bad" interpolant is used for the analysis. When a good interpolant is used, the gains are relatively modest. This is again because the loss of contrast has a larger influence on the results than the added noise of the aliased information. This study does confirm that having an aliased speckle is much worse (approximately 3×) than having a larger and fully resolved speckle. Therefore, aliased speckles should always be avoided. Finally, with modern DIC software, it seems that adding an anti-aliasing filter to commercial machine vision cameras is not worth the added cost and effort.

**Acknowledgements** I would like to thank my summer intern, Hailey Stock for acquiring this data, and Paul Farias for installing the anti-aliasing filters on the cameras. Sandia is a multiprogram laboratory operated by Sandia Corporation, a Lockheed Martin Company, for the United States Department of Energy's National Nuclear Security Administration under contract No. DE-AC04-94AL85000.

## References

1. Sutton, D.A., Orteu, J.J., Schreier, H.W.: Image Correlation for Shape, Motion and Deformation Measurements. Springer, New York (2009)
2. Schreier, H.W., Braasch, J.R., Sutton, M.A.: Systematic errors in digital image correlation caused by intensity interpolation. Opt. Eng. **39**(11), 2915–2921 (2000)

# Chapter 15
# Investigation of Electronic Speckle Pattern Interferometry with Line Laser Scanning for Large Area Deformation Measurement

**Shuichi Arikawa and Yuta Ando**

**Abstract** For measuring small deformations in large structures by electronic speckle pattern interferometry (ESPI), the increase of laser power is required for providing the sufficient laser power per area. However, high-power lasers lead increasing the risk of exposure to laser, the size of the equipment and the cost. In this study, ESPI with line laser scanning is investigated for large area deformation measurements without increasing the laser power. A dual-beam interferometer for a horizontal displacement measurement which can illuminate horizontal line lasers and can scanning for vertical direction is constructed. A static in-plane deformation of an aluminum sheet specimen is measured. Speckle images for each vertical position are captured at the initial and the deformed state. Analyzed phase maps for each line show mismatch of the phases. Additionally, a method for integrating the mismatched line phase maps are investigated.

**Keywords** Deformation Measurement • Speckle Interferometry • Line Laser • Laser Scanning • Phase Integration

## 15.1 Introduction

For measuring surface deformations, optical or interferometric techniques are effective. Especially, electronic speckle pattern interferometry (ESPI) [1, 2] is very useful for measuring elastic deformations or thermal deformations because of the high sensitivity. When large area is measured by ESPI, a high power laser source is required to ensure the sufficient laser power per area. It is expected that a very high power laser as watt level is required for meter order of the measurement area. High power lasers more than 0.5 watt is classified class 4 as the most dangerous level. It leads requirements of safety equipment and increasing the cost. Therefore, the development of the technique for large area deformation measurement by ESPI without increasing the laser power is effective for the usefulness. In this study, ESPI with line laser scanning for large area deformation measurements are investigated.

## 15.2 Electronic Speckle Pattern Interferometry with Line Laser Scanning

A schematic figure of a dual-beam interferometer for ESPI with line laser scanning is shown as Fig. 15.1. The line shaped interference region is made on the object surface using cylindrical lenses and can scan the object surface by moving the interferometer. Image capturing at the same scanning positions before and after the deformation are required. Additionally, an investigation of the method for integrating the line interferograms. Figure 15.2 shows a schematic of the image integration. Phase mismatches occur by environmental vibrations. Therefore, the phase correction for the image integration is necessary. In this study, phase values in overlap regions arranged between adjacent lines are compared. The line phases are then corrected and integrated.

S. Arikawa (✉) • Y. Ando
Department of Mechanical Engineering Informatics, School of Science and Technology, Meiji University, 1-1-1 Higashimita, Tama-ku, Kawasaki, Kanagawa, 214-8571, Japan
e-mail: arikawa@meiji.ac.jp



L. Lamberti et al. (eds.), *Advancement of Optical Methods in Experimental Mechanics, Volume 3*,
Conference Proceedings of the Society for Experimental Mechanics Series, DOI 10.1007/978-3-319-63028-1_15

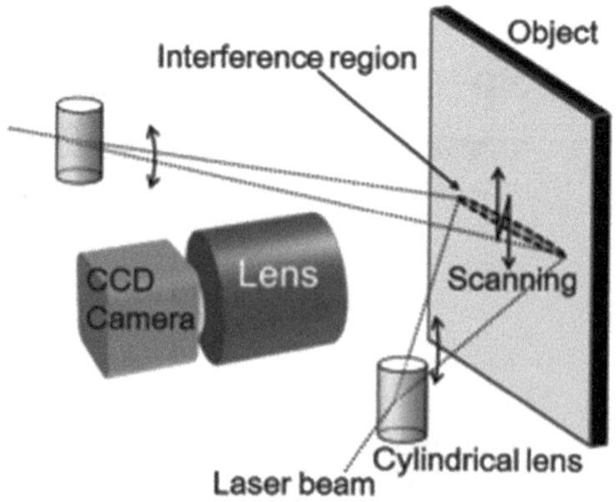


**Fig. 15.1** Schematic figure of line laser scanning speckle interferometry

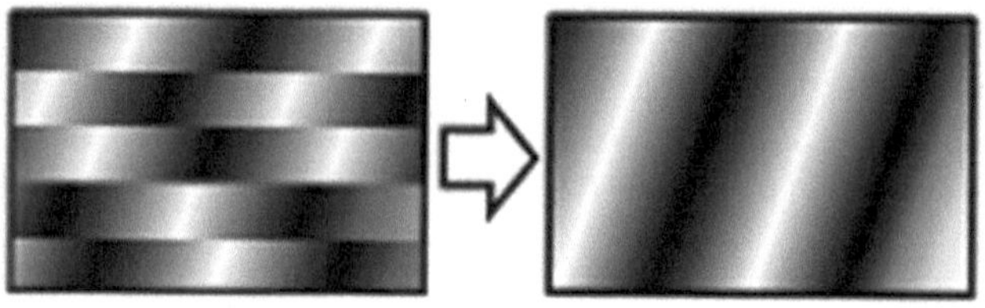

**Fig. 15.2** Line phase integration

## 15.3 Experiment

A dual-beam interferometer for a horizontal displacement measurement which can illuminate horizontal line lasers and can scanning for vertical direction is constructed on an experimental stage. The laser power is 53 mW and the wavelength is 633 nm. The beam diameter for vertical axis is 1.42 mm. The width of the interference region expanded by the cylindrical lenses is about 50 mm. A monochrome CCD camera is used for measurement. A measurement of a static horizontal tensile deformation of an aluminum sheet specimen is carried out. The camera is fixed on the experimental stage while the measurement with laser scanning.

## 15.4 Results and Discussion

Line shaped interferograms showing the tensile deformation were obtained. Hence the accurate laser scanning could be performed. Analyzed phase maps for each line showed the mismatch of the phases. The phase correction was performed. The corrected phases were then integrated as a single phase map. The integrated phase map showed a continuous phase distribution. Therefore the phase correction was effective. However, slight phase mismatches observed on the integrated phase map with phase correction when the scanning steps increased. It is expected that the deformation slightly occurs while the measurement. The proposed method can only be used for static deformation measurement. However, a comparatively long time was taken for the line laser scanning and the phase shift for the phase analysis because of the hand movement at this stage. Therefore, the deformation increased during the measurement. For minimizing the deformation during the measurement, speeding up the measurement process by an automation is required.

## 15.5 Conclusion

In this study, ESPI with line laser scanning was investigated for large area deformation measurements without increasing the laser power. A dual-beam interferometer for a horizontal displacement measurement which can illuminate horizontal line lasers and can scanning for vertical direction was constructed. A static in-plane deformation of an aluminum sheet specimen was measured. Speckle images for each vertical position were captured at the initial and the deformed state. Analyzed mismatch phases were corrected and integrated. As the result, accurate laser scanning could be performed. The continuous phase distribution was obtained by the image integration with the phase correction. Therefore, the proposed method is effective. Additionally, for increasing the scanning steps, speeding up the measurement process by an automation is necessary.

## Reference


1. Chiang, F.P.: Speckle Metrology, ASM Handbook Volume 17, Nondestructive Evaluation and Quality Control, pp. 432–437. ASM International, OH, USA (1989)
2. Sirohi, R.S.: Speckle interferometry. Contemp. Phys. **43**(3), 161–180 (2002)


**Shuichi Arikawa** Degree: Doctor of Engineering from Yokohama National University; Position: Senior Assistant Professor; Affiliation: Department of Mechanical Engineering Informatics, School of Science and Technology, Meiji University; Interest: Material science, Solid mechanics, Optical measurement technique.

# Chapter 16
# Internal Heat Generation in Tension Tests of AISI 316 Using Full-Field Temperature and Strain Measurements

**Jarrod L. Smith, Veli-Tapani Kuokkala, Jeremy D. Seidt, and Amos Gilat**

**Abstract** Full-field temperature and strain measurements were recorded during tension tests of AISI 316 on a hydraulic load frame at a strain rate of 1 $s^{-1}$. The temperature increase was measured on one side of the specimen using a high speed IR camera while the deformation was measured on the opposite side with a visible camera, each at a frame rate of 500 FPS. Uniform deformation of the specimen was observed up to strains of 0.25 until necking occurred and localization strains reached up to 0.75 at failure. The maximum temperature as measured by the IR camera was 260 °C before failure. The fraction of plastic work converted to heat (ß) was calculated over the entire gage length of the specimen using the local measurements of stress, strain, and temperature and varied between 0.6 and 0.9 throughout the test.

**Keywords** Infrared Thermography • Digital Image Correlation • Plastic Work • Thermomechanical • Stored Energy

## 16.1 Introduction

Inelastic deformation behavior of all metals is known to be affected by strain rate and temperature. The effects of strain rate and temperature are coupled because a significant portion of the energy required for deformation is also converted to heat during the process which raises the temperature of the material. At low strain rates, plastic deformation takes place in virtually isothermal conditions because as the heat is generated in the material, there is sufficient time for the heat to dissipate into the surroundings. At higher strain rates the temperature in the material reaches higher temperatures because the event occurs in period of time that is too short for the heat to transfer from the material. To observe this behavior, material specimens are normally tested in simple tension tests utilizing a load frame for quasi-static tests and Split Hopkinson Bar for higher strain rate tests. The force on the specimen can be measured by utilizing a load cell for quasi-static tests and by analyzing the elastic waves in the Split Hopkinson Bar. The strains on the surface of the specimen can be determined utilizing Digital Image Coorelation (DIC). DIC is implemented by painting the specimen surface with a black and white speckle pattern and capturing the deformation of the surface with high speed cameras. The images are then processed with software that determines the displacements of the speckles during deformation and then calculates the strains based off of the displacements. Due to recent advancements in Infrared Thermography cameras it is also possible to measure the temperature on the surface of the specimen at high speeds. Simultaneous full-field measurement of the strain and temperature has been completed by many researchers at low strain rates. For example, Saai *et al.* [1] studied the thermomechanical behavior of Al bi-crystals with tensile tests at the strain rate of ca. $10^{-2}$ $s^{-1}$ making use of simultaneous DIC and IR measurements. In their experiments, the infrared and visible cameras were observing the same specimen area, which was first coated with black paint and then sprayed with white paint to form a random pattern for DIC. They estimated that the emissivity of the paint(s) was 0.96, i.e., very close to that of a black body. The IR images were recorded during the tensile tests at 20 frames per second (fps) with a resolution of 320 × 240 pixels. Oliferuk *et al.* [2] used infrared thermography and visible imaging to determine the energy storage rate in the area of strain localization in an austenitic stainless steel similar to AISI 304 L. In their tests, the mean value of strain rate was $6.6 \times 10^{-1}$ $s^{-1}$ and the imaging frame rate of both visible and IR cameras was 538 fps. For local strain determinations, graphite dot markers were painted on one surface of the specimen, while the opposite surface on the IR camera side was covered with soot with estimated emissivity of 0.95.

J.L. Smith (✉) • J.D. Seidt • A. Gilat
Department of Mechanical Engineering, The Ohio State University, Scott Laboratory, 201 W 19th Ave, Columbus, OH, 43210, USA
e-mail: smith.6575@osu.edu

V.-T. Kuokkala
Department of Materials Science, Tampere University of Technology, P.O.B 589, 33101, Tampere, Finland



L. Lamberti et al. (eds.), *Advancement of Optical Methods in Experimental Mechanics, Volume 3*,
Conference Proceedings of the Society for Experimental Mechanics Series, DOI 10.1007/978-3-319-63028-1_16

The calculation of the fraction of mechanical energy converted to thermal energy during this test can be calculated as we have synchronous measurements for force, strain and temperature. This fraction of plastic work dissipated as heat is usually denoted by ß and has been theorized and calculated by researchers in the past. Rosakis et al. [3] first presented the theoretical thermodynamic foundations illustrating the plastic strain and strain rate dependence of ß. Hodowany et al. [4] then expanded on these tests by measuring the temperature of Al2024-T3 and α-Titanium at strain rates of 1 $s^{-1}$ and 3000 $s^{-1}$. From these test he concluded that the Al2024-T3 had a plastic strain but not strain rate dependent ß while the ß for α-Titanium was both plastic strain and strain rate dependent. ß was calculated in the preceding tests with the following equation: $\beta = \frac{\rho c \dot{\theta}}{\sigma \dot{\varepsilon}^p}$ where $\rho$ is the mass density, c is the specific heat, $\dot{\theta}$ is the change in temperature, $\sigma$ is the component of stress and $\dot{\varepsilon}^p$ is the change in plastic strain.

In the present paper, simultaneous full-field strain and temperature measurements, using 2D DIC and a high-speed IR camera, respectively, are made in tensile tests at a strain rate of 1 $s^{-1}$. The results show the evolution of strain and temperature during the uniform deformation in the early part of the test and in the necking region during the localization. Calculations of the partition of plastic work converted to heat based on the measurements of force, strain, and temperature are presented. Practical and theoretical issues related to the emissivity of the specimens, calibration of infrared cameras, as well as the current limitations of the infrared techniques, such as the limited temperature range of measurements at a constant exposure time are discussed.

## 16.2 Experimental Procedures and Techniques

Tensile tests on flat thin stainless steel 316 specimens are conducted at a strain rate of 1 $s^{-1}$. Full-field deformation and full-field temperature are measured simultaneously and synchronously during the tests. The deformation is measured using DIC on one side of the specimen, and the temperature is measured on the opposite side with an IR camera. Since the specimen is thin, it is assumed that the deformation and temperature are uniform through the thickness. The quasi-static test is completed using a servo-hydraulic MTS load frame as shown in Fig. 16.1 below.

### *16.2.1 Specimen Geometry and Material*

Tests at high strain rates require specimens with a short gage length, and in order to eliminate possible effects of the specimen geometry on the results, the gage section geometry of the specimens used in the quasi-static tests is the same. A drawing of the specimen used in the high strain rate tests is shown in Fig. 16.2. The specimens are made of stainless steel AISI 316.

**Fig. 16.1** Testing configuration on the MTS frame

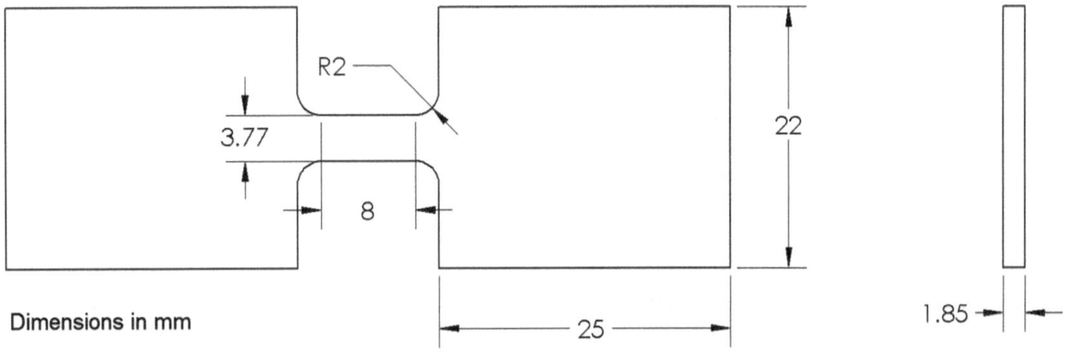


**Fig. 16.2** Specimen Geometry

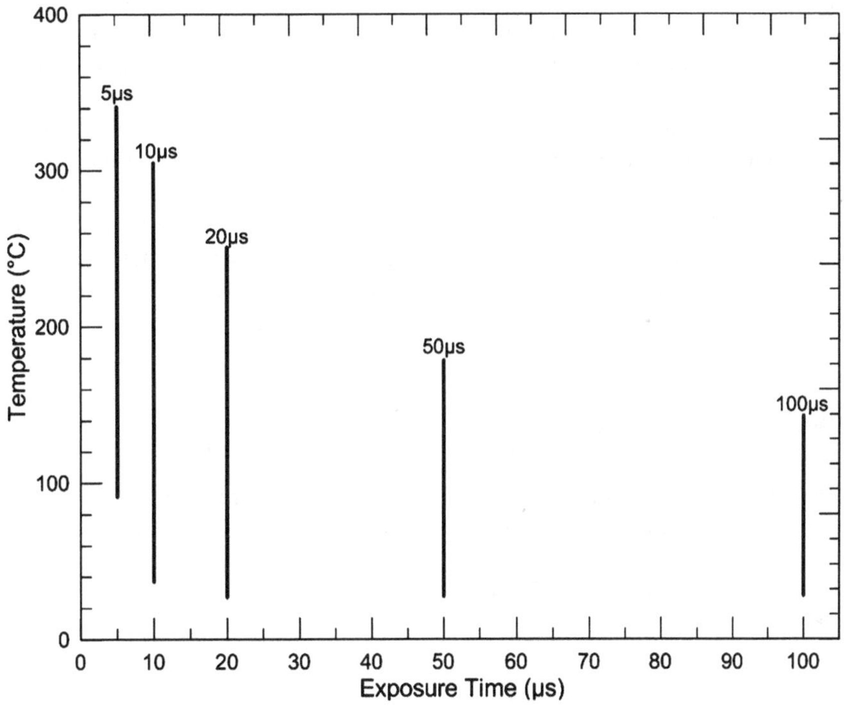


**Fig. 16.3** Exposure time *vs.* temperature range of the Telops high speed infrared camera

This material exhibits high ductility, strong capability of strain hardening, and low thermal conductivity, which all together lead to notable internal heating of the material already at relatively low strain rates.

### *16.2.2 Full-Field Thermal Imaging and Calibration*

Thermal image capture is done with Telops FAST-IR 1000 high speed infrared camera, capable of 256 × 320 pixel resolution at speeds up to 1,000 fps and a reduced pixel resolution at speeds up to 30,000 fps. The exposure time of the IR camera for the specified test was 10 μs. With decreasing exposure time the IR camera can register higher surface temperatures, but at the same time the lowest measurable temperature increases so that the usable temperature range at each constant exposure time is ca. 250 K, as seen in Fig. 16.3.

To synchronize the Phantom visible camera and the IR camera, they are externally driven by a dual-channel Keysight 33500 series waveform generator, which enables running the cameras at different but phase-synchronized frequencies. In practice, the visual camera typically is run at a frequency which is the same or a multiple of the IR camera frequency. The waveform generator sends TTL level timing pulses to both cameras in an externally triggered burst mode. In tests with the MTS frame, the external trigger to the waveform generator is sent through the digital output port of the controller unit 100 ms before the actuator starts to move. The data acquisition of the MTS controller starts at the same time as the trigger is sent, and therefore the images taken by both cameras as well as the load cell and LVDT signals of the load frame have a common recorded time base.

The infrared camera used in this work can record the temperature data as so-called radiometric temperatures or as non-uniformity corrected photon counts for each pixel. Since the radiometric temperature method assumes that the emissivity of the object is unity, i.e., it is a black body, the temperature readings obtained for stainless steel specimens will be too low because of its lower emissivity. Emissivity values for various substances can be found in the literature, but since the actual emissivity depends on many factors, such as the surface finish of the object, a calibration with real samples is often the only way to obtain reliable temperature values. As stated in the Introduction, in many previous works a coating, such as soot or spray paint, has been applied on the specimen surface to increase its emissivity close to that of a black body. The downside of this approach is that the coating tends to peel off, crack, or thin too much especially in the necking area of the specimen at large strains, which increases the uncertainty of the obtained temperature readings. In this work, calibration measurements with several specimens were conducted in the following manner: thermocouples were attached to the surface of a specimen, which was heated on a hot plate in 10–20 K steps and the true temperature, indicated by the thermocouples, was recorded together with the IR camera reading obtained as an average of an area of reasonable size. To examine the effect of plastic deformation on the emissivity the calibration was done with tested specimens which have areas with different amount of plastic deformation. The necking area is highly deformed and appears a little rougher compared to the surface outside the gage section which is not deformed plastically. In the calibration procedure, IR camera readings at all temperature steps from room temperature up to ca. 350 °C were recorded from the deformed sections of tested specimens using exposure times of 100 μs, 50 μs, 20 μs, 10 μs, 5 μs and 2 μs. From the calibration measurements, polynomial fits were formed separately for each exposure time, which were then used to convert the radiometric temperatures given by the IR camera to true surface temperatures of the specimen. The temperatures were then interpolated at the points where strains were calculated on the surface and using these synchronous measurements the partition of plastic work converted to heat at each point was calculated at each point of strain.

## 16.3 Experimental Results

The measured true strain and temperature at the point of maximum strain at failure in the specimen is displayed in Fig. 16.4. The strain measured by a 4 mm extensometer over the gage length of the specimen is also shown. The two values of strain are comparable until the onset of localization at a strain of approximately 0.3. It is at this point that a large increase of temperature is also observed the peak value of 250 °C is reached at a local strain of 0.8.

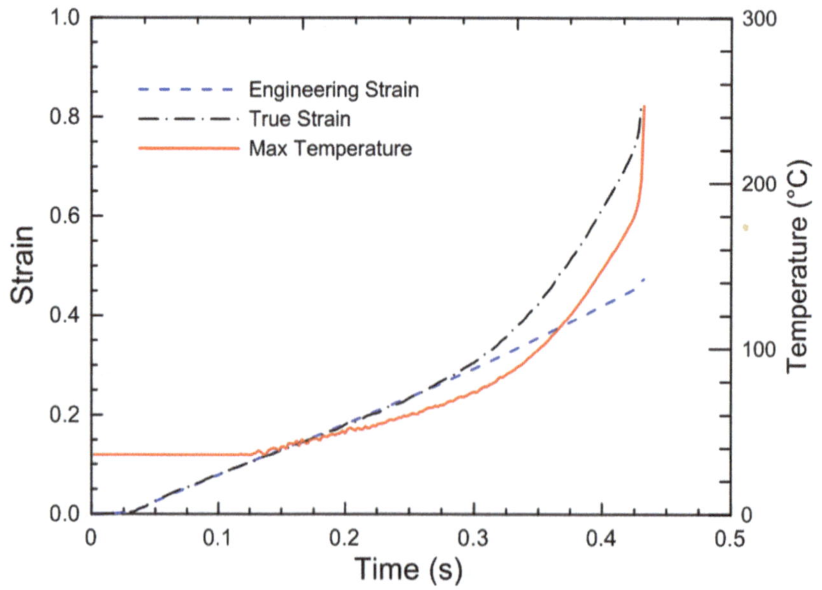


**Fig. 16.4** History of strain and temperature at the point of maximum strain

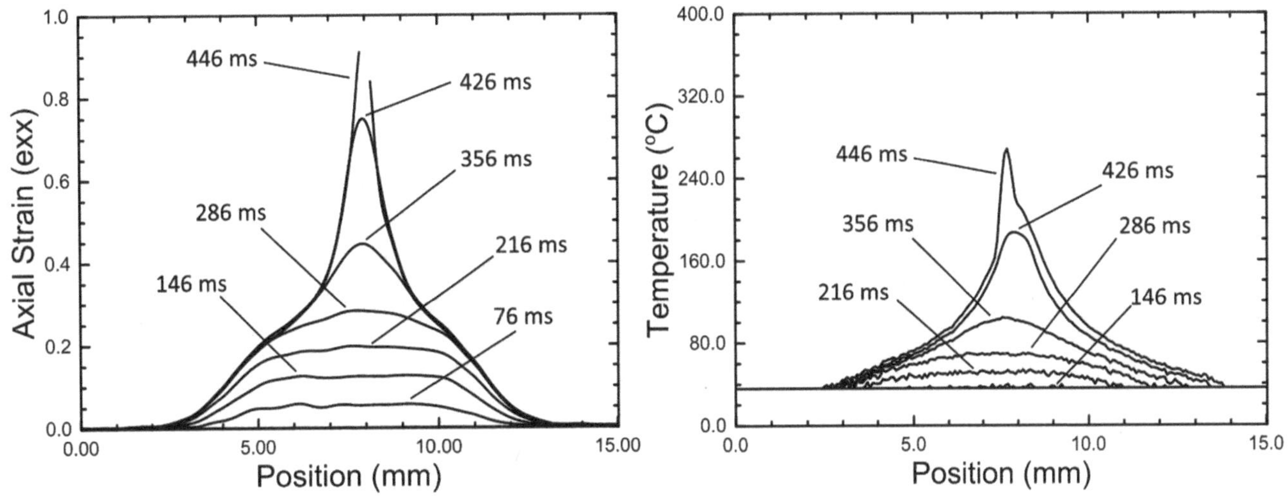


**Fig. 16.5** Strain and temperature along the specimen at various times

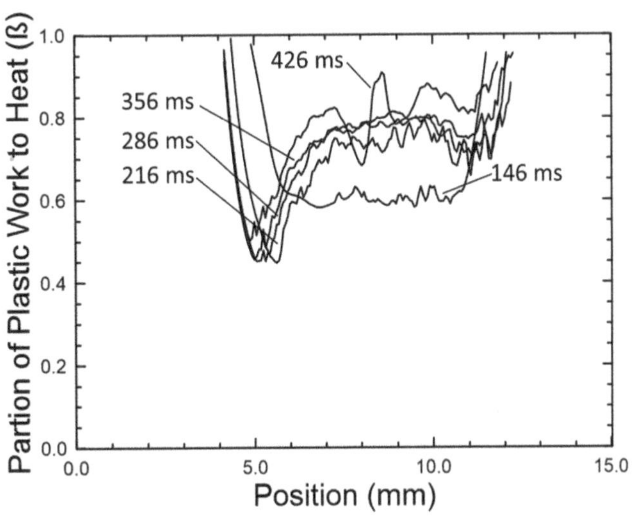


**Fig. 16.6** Calculated beta along the specimen at various times

In Fig. 16.5, the rise in temperature is characterized with the rise in axial strain across the length of the specimen. The strain and temperature measurements along the gage are almost uniform along the gage section up until 356 ms into the test. The large increase in the localization temperature and strain are also evident in these graphs, reaching their peak levels at the point before failure (426 ms) and directly after (446 ms). The measured strains and temperatures in Fig. 16.5 were then used in concert with the estimated stresses in the specimen to calculate the partition of plastic work (ß) converted to heat over the gage length of the material. It is evident that during the early stages of deformation the strain energy is mostly being transferred to the mechanical work in the material, resulting in a beta lose to 0.6 over the length of the gage section. However, as the deformation increases the beta to 0.8–0.9 across the specimen as the temperature of the material rises and more energy is released as heat (Fig. 16.6).

Figure 16.7 shows the calculated beta at the maximum point of strain from Fig. 16.4 as well as the estimated temperature based on the measured stress and strain data coupled with a constant beta value. As was displayed in the waterfall plots, the beta value is highly dependent on the level of strain in the material as the beta ranges from 0.6 to 0.9 throughout the test. The average beta for this point was calculated as 0.786.

Figure 16.8 shows the large amount of data that is generated in a single test as 24,000 separate points have measurements of strain, interpolated temperatures, and calculated betas.

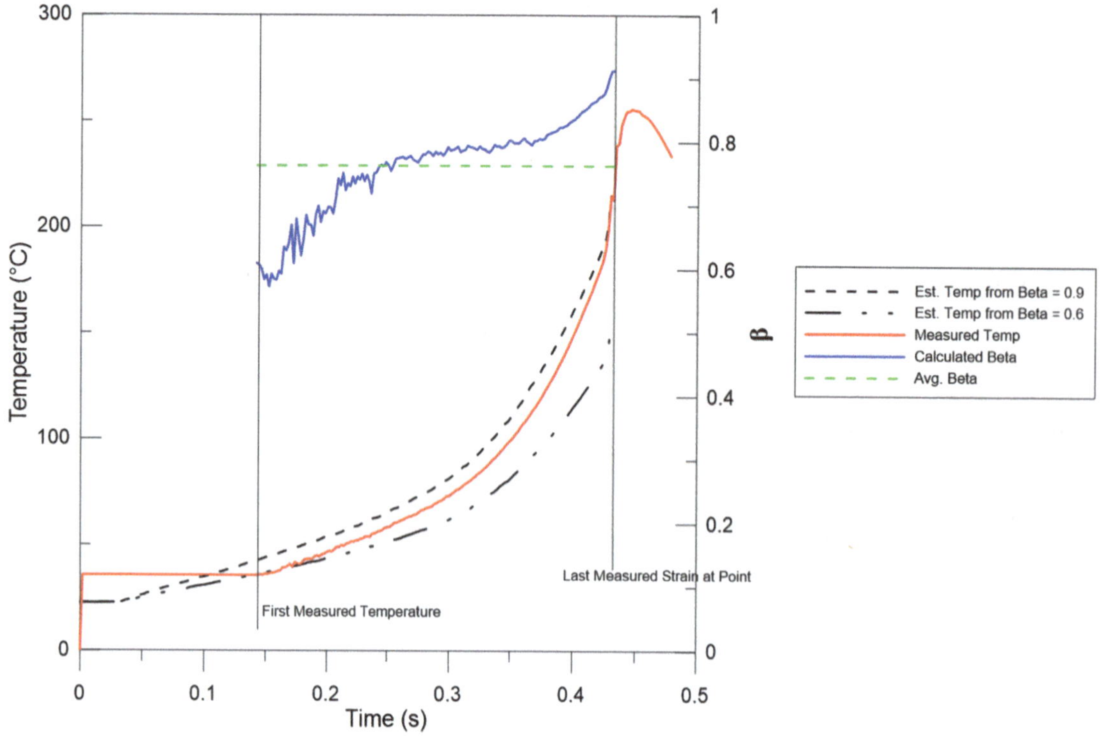


**Fig. 16.7** Calculated beta and measured temperature at point of maximum strain versus estimated temperature from constant beta values

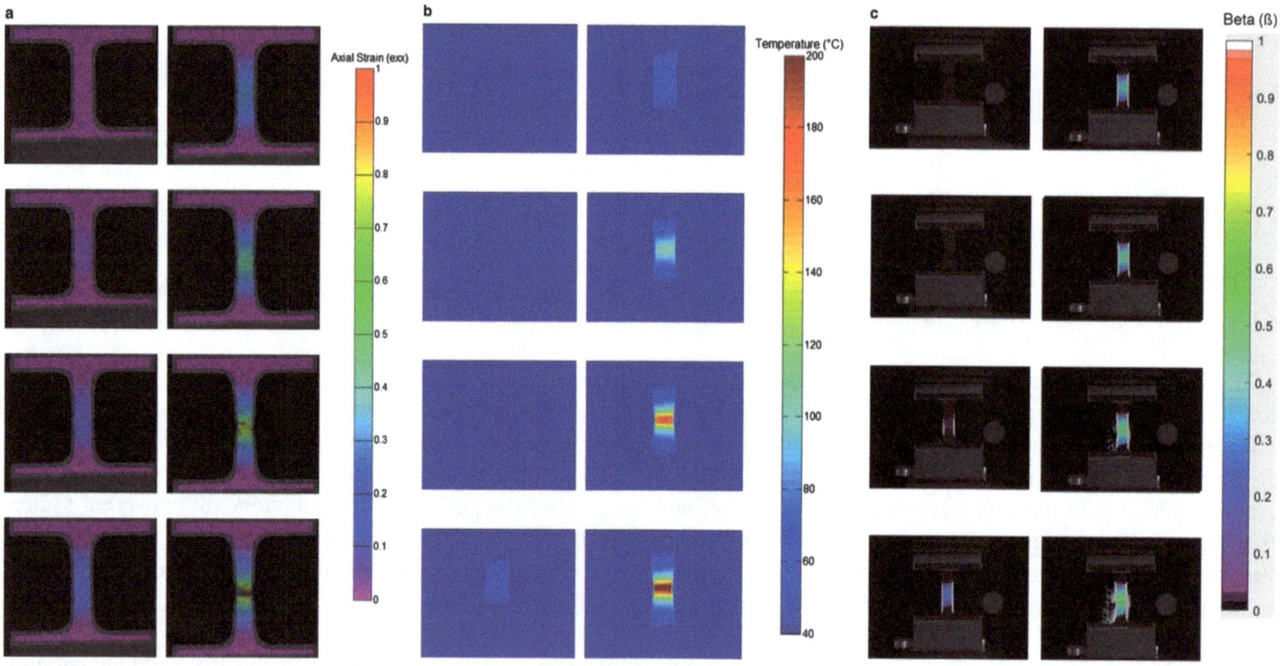


**Fig. 16.8** Full field plots of strain, temperature and beta

## 16.4 Summary and Conclusions

Simultaneous full-field deformation measurements, temperature measurements, and beta calculations on the surface of the specimen have been completed in a tension test at 1 $s^{-1}$. There is uniform temperature rise during the uniform deformation before necking and faster localized deformation and heating in the neck region during necking. The partition of plastic work converted to heat increases as the plastic strain in the specimen increases. Similar tests at higher strain rates should be conducted to determine if the beta is also strain rate dependent.

**Acknowledgements** The research was supported by the U.S.A. Federal Aviation Administration, Grant No. 11-G-004. The authors are grateful to Mr. William Emmerling, and Dr. Chip Queitzsch for their support and involvement.

## References

1. Saai, A., Louche, H., Tabourot, L., Change, H.J.: Experimental and numerical study of the thermo-mechanical behavior of al bi-crystal in tension using full field measurements and micromechanical modeling. Mech. Mater. **42**(3), 275–292 (2010)
2. Oliferuk, W., Maj, M., Zembrycki, K.: Determination of the energy storage rate distribution in the area of strain localization using infrared and visible imaging. Exp. Mech. **55**(4), 753–760 (2013)
3. Rosakis, P., Rosakis, A.J., Ravichandran, G., Hodowany, J.: A thermodynamic internal variable model for the partition of plastic work into heat and stored energy in metals. J. Mech. Phys. Solids. **48**(3), 581–607 (2000)
4. Hodowany, J., Ravichandran, G., Rosakis, A., Roaskis, P.: Partition of plastic work into heat and stored energy in metals. Exp. Mech. **40**(2), 113–123 (2000)

**Jarrod L. Smith** is a 2nd year PhD student at the Ohio State University. In the past Jarrod has focused his research on the full-field strain and infrared radiation measurements of metals at various strain rates. The paper Jarrod is presenting today is the beginning of his dissertation work on the fraction of work that is converted to heat.

# Chapter 17
# A Short Survey on Residual Stress Measurements by HDM and ESPI

**C. Pappalettere**

**Abstract** In this paper a review of progresses occurred along the years in measuring residual stresses by optical methods is presented. These methods allow to implement the hole drilling procedure for residual stress profile measuring, without applying the strain gage rosette. This approach presents several advantages such as easier and cheaper preparation of the test beside the possibility to avoid eccentricity issues and to increase the amount of available data in view of the fact that each pixel can be considered as a measure point. In particular the evolution of the Electronic Speckle Pattern Interferometry (ESPI) technique will be presented showing how set up, methodologies and calculation approaches have developed along the last three decades.

**Keywords** ESPI • Hole drilling method • Residual stresses • Optical methods • Interferometry

## 17.1 Introduction

The stress field existing in some materials without application of an external source of stress, such as loads or thermal gradients, is known as residual stress. These residual stresses (RS) are generated in almost all manufacturing processes such as machining, grinding, forming [1], rolling, casting, forging [2], welding [3–5], melting [6], cladding [7] heat treatment [8], etc. or may occur during the life of structures. The hole drilling method (HDM) is one of the mo st widely used techniques for measuring residual stresses. This technique consists in the localized removal of stressed material and in measuring the strain field consequent to the relieved stresses. The hole drilling method using strain gauge rosettes is a consolidated approach for stress determination and it follows the ASTM test standard. Even though strain gauges are usually used to measure these displacements, they have some disadvantages: the specimen surface has to be flat and smooth so that the rosettes can be attached, the surface of the material has to be accurately prepared, the hole has to be drilled exactly in the center of the rosette in order to avoid eccentricity errors, and time and costs associated with installing rosettes are consistent. Furthermore the amount of available data is limited: for each measurement, only three discrete readings are available (six in the case of some special rosettes), just sufficient to fully characterize the in-plane residual stresses. Due these considerations several attempts to replace extensimetric measurements by optical ones have been done along the years. For example the use of moiré interferometry for strain determination in RS measurements by HDM was investigated in many situations since McDonach et al. [9]. However, bonding a grating can also be time consuming. The feasibility of using holographic interferometry was shown by Antonov [10]. More recently Hung et al. [11] have used shearography in conjunction with a small ball indentation to determine RS while in [12] shearographic and speckle approach are compared. Also the possibility to release residual stresses by using local heat treatment, combined with electronic speckle pattern interferometry was attempted in [13–15] and more recently repeated on aluminum alloys in [16, 17]. However the most explored approach consists in combining ESPI with incremental hole drilling; the basic idea is to adopt the same procedure for RS measurement as described in ASTM E837 but replacing data obtained by the rosette with data obtained by speckle pattern. Speckle interferometry approach is widely adopted in experimental mechanics [18–23]. It is based upon the observation that whenever a coherent beam of light shines on a surface a random intensity pattern is produced as a consequence of the interference of the light diffused by surface itself. Displacements of the surface introduce changes in the local phase. Having this in mind it is easy to understand that if an hole is drilled in a material and stresses are relaxed speckle patterns change and if the speckle pattern is subtracted

C. Pappalettere (✉)
Politecnico di Bari, Dipartimento di Meccanica, Matematica e Management, Viale Japigia 182, 70126, Bari, Italy
e-mail: c.pappalettere@poliba.it



L. Lamberti et al. (eds.), *Advancement of Optical Methods in Experimental Mechanics, Volume 3*,
Conference Proceedings of the Society for Experimental Mechanics Series, DOI 10.1007/978-3-319-63028-1_17

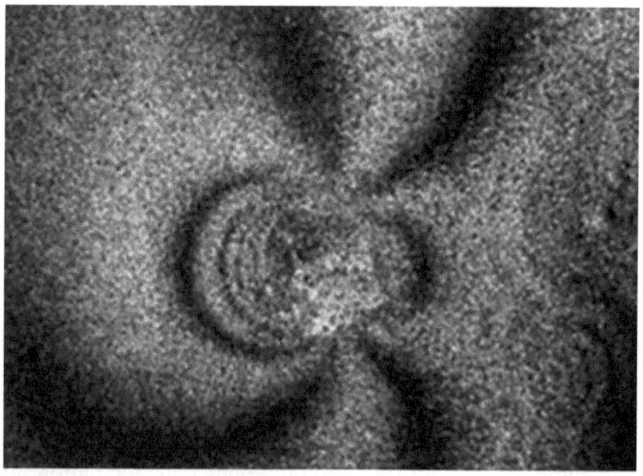

**Fig. 17.1** Example of ESPI correlation fringes obtained during the hole-drilling process. Fringes are obtained by subtracting the speckle pattern recorded at a given drilling increment from the reference pattern recorded before starting the drilling procedure. Fringes represent the loci of points of isodisplacement along the direction of the sensitivity vector

from a reference pattern correlation fringes are obtained as those shown in Fig. 17.1. In this paper the evolution of the ESPI-HDM technique along the years is presented showing that systems have become increasingly compact and easy to operate, calculation methods have become more robust and faster and the field of application of the technique has increased more and more.

## 17.2 Origin of ESPI-HDM

Pioneering works in the adoption of ESPI combined with Hole Drilling methods are those of Furgiuele et al. [24]. The optical set-up adopted allowed to determine three components of the displacement in the plane of the specimen and two oblique to the plane. In order to verify the reliability of the method the system was tested on a steel specimen on which a well know stress state was introduced. The centers of the fringes were detected manually that is to say that it was a task of the user to track them on the images and to assign them the proper order of fringe. This allowed to determine the displacement on the pixel along the fringes and to obtain stresses based upon calibration coefficients connecting the stress state to the displacement occurring as a consequence of the drilling procedure. In spite of the complexity of the optical configuration adopted and of the difficulties in speckle pattern elaboration authors obtained very promising results that paved the way to further developments. Analogous results, in terms of accuracy, were obtained 2 years later in [25] where, instead of a CCD camera as a recording media, photographic emulsions were adopted (Fig. 17.2).

## 17.3 Development of the Technique

Attempts to simplify the optical system with respect to pioneering works were firstly done in [26] where a system based upon a couple of optical fibers was used to implement a double Lendeertz configuration showing potentialities of this technique to be implemented in a easy to operate set-up. A similar system will be also used later for industrial application such as RS measurement in CD disc [27]. A simplified set up was also adopted later where an ESPI out-of-plane interferometer was built however, in that case, the chemical etching procedure was used instead of hole drilling to get stress relaxation along the depth. In the direction of implementing the technique in a robust a portable device it is noticeable the work done in [28], in this paper by the use of a conical mirror, an optical configuration was obtained so that the sensitivity vector is directed along the radius for each point of the analyzed specimen. The system was also tested on the field by measuring residual stresses on a pressure vessel while the same interferometric configuration was adopted in combination with indentation instead of hole drilling in [29] even if a numerical model for quantitative evaluation of RS is missing. Very huge work from the side of system development, data analysis development and automation, error sources evaluation and results validation was done

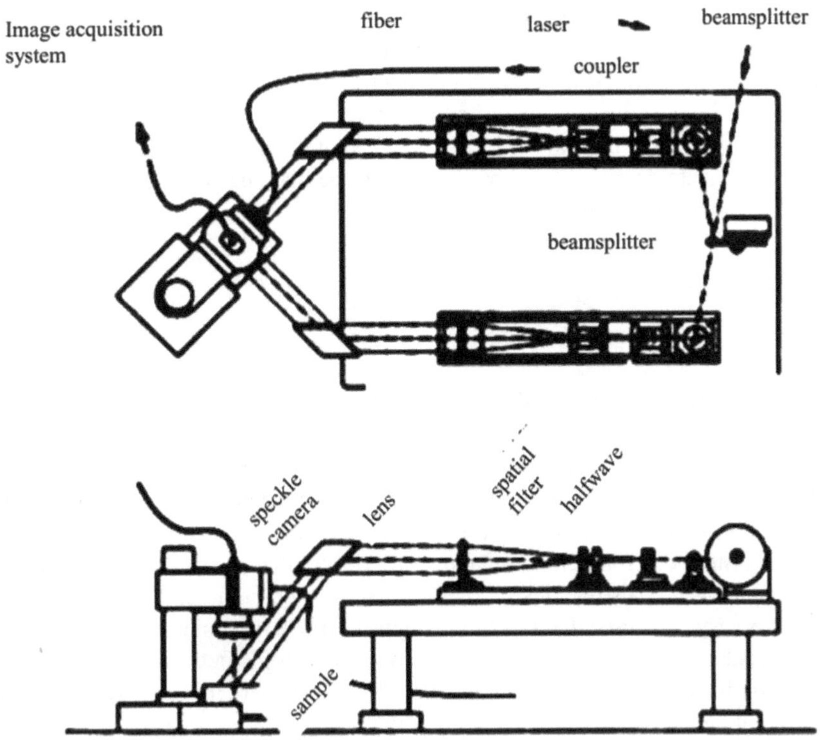


**Fig. 17.2** Set-up for HDM-ESPI adopted in the pioneering work by Furgiuele et al. [24]

in the series of papers [30–34] as well as in [35] while in [36] repeatability of stress measurement in ESPI hole-drilling was studied. The system described in those papers is quite compact; the sample is illuminated just from one side by an optical fiber while a reference beam is sent directly to the camera CCD by a second optical fiber. A database of finite element solutions was built allowing to determine surface deformation for each combination of Young's modulus and Poisson's ratio, hole geometry and stress state. Full field least square technique was implemented to determine the stress state as the combination of σx, σy and τxy that minimizes the differences between the measured displacements and the modelled ones. The algorithm allowed also to take into account the rigid body motion that can occur during the measurement. Furthermore some sources of errors connected with the measurement such as the error on the determination of the illumination and observation angles or in the determination of the pixel size are taken into account. Further work in this sense has be done later on in [37] where effects of drilling parameters are considered for aluminum and stainless steel or in [38–43] where analogous analysis is carried out for Titanium grade 5. In [44–46] other sources of errors are considered such as the extension of the analysis area that is to say the effective region of the acquired interferogram that will be used for the analysis. In [48] was explored, with a good success, the possibility to perform remote drilling operation that means to afford issues connected with replacement of the sample and consequent fringe decorrelation. The issue was overcome by using a kinematic mount with two pins and two bolts and by implementing rigid body correction algorithms. Optimization of the entire measurement processes are analyzed in [49] by considering the problem of full exploitation of all available data and in [50] by considering the problem of the choice of the number and distribution of drilling steps.

## 17.4 Recent Progresses

In the last decade the number of works where hole drilling method an ESPI are used in combination to measure residual stresses has increased very much with applications ranging from measurement on submerged arc welded parts [51], TIG welded structural elements [52], FHPP welded part [53], butt-welded joints [55], dissimilar-weld joint [55] 3D printed part by Fused Deposition Modelling [56–58], glass and plastics [59, 60] etc.. At the same time some alternatives approaches also started to be attempted, for example in [61] incremental grooving was used instead than drilling while in [62–63] cross slitting and dual-axis ESPI were introduced demonstrating that this configuration gives better results with respect to the determination of the shear stress profile. In [64] an hybrid approach combining hole-drilling and ring-core techniques was studied. Most recent works are devoted to the improvement of calculation methods; in [65] the integral method for stress calculation in incremental hole drilling was modified so that the speckle image recorded at each drill increment is not correlated with the initial reference state but with the image obtained at the previous step. The matrix that rules the problem in such a case connecting incremental strain to stresses turns out to be diagonally dominant so that all the system is better conditioned and error amplification is reduced.

## 17.5 Final Considerations

Literature review clearly displays how the use of ESPI in combination with hole-drilling method has greatly developed along the years. Technological evolution in terms of smaller, cheaper and higher resolution CCD camera as well as smaller and higher quality laser sources allows now to obtain very robust and compact setup that can also be used outside the laboratory [66–67]. At the same time calculation algorithm have been refined along the years and a series of tools have been introduced so that fringe analysis can be considered to be a semi-automated process that can be quite easily managed also by a trained but not expert operator even in industrial environment as it happens for different approaches [68, 69]. Due to these considerations technology appears to be mature as it is also witnessed by the presence of commercial systems available on the market and based upon this approach. It is easy to foresee that this kind of approach will be subjected to further development and spread. At the same way it can be speculated that new research will be done on the topic in the next years for example in the direction to extend its application to non-isotropic materials or in the direction to compare this approach with a more recent one based upon detection of displacement fields by Digital Image correlation [70–73].

## References


1. Lee, K.T., Park, C.S., Kim, H.Y.: Fatigue and buckling analysis of automotive components considering forming and welding effect. Int. J. Automot. Technol. **18**(1), 97–102 (2016)
2. Fu, Y., Li, W.Y., Yang, X.W., Ma, T.J., Vairis, A.: The effects of forging pressure and temperature field on residual stresses in linear friction welded Ti6Al4V joints. Adv. Manuf. **4**(4), 314–321 (2016)
3. Casavola, C., Lamberti, L., Pappalettere, C., Tattoli, F.: A comprehensive numerical stress – strain analysis of laser beam butt-welded titanium compared with austenitic steel joints. J. Strain Anal. Eng. Des. **45**(7), 535–554 (2010)
4. Casavola, C., Pappalettere, C.: Discussion on local approaches for the fatigue design of welded joints. Int. J. Fatigue. **31**(1), 41–49 (2009)
5. Casavola, C., Pappalettere, C.: Application of WEL.FA.RE. method on aluminum alloy welded joints. In: Proceedings of the 2005 SEM Annual Conference and Exposition on Experimental and Applied Mechanics, pp. 1555–1562 (2005)
6. Casavola, C., Campanelli S., Pappalettere, C.: Experimental analysis of residual stresses in the selective laser melting process. Society for Experimental Mechanics – 11th International Congress and Exhibition on Experimental and Applied Mechanics 2008, vol. 3, pp. 1479–1486 (2008)
7. Alam, M.K., Edrisy, A., Urbanic, J., Pineault, J.: Microhardness and stress analysos pf laser-cladded AISI 420 martensitic stainless steel. J. Mater. Eng. Perform. **1**, 1–9 (2017)
8. Araghchi, M., Mansouri, H., Vafaei, R., Guo, Y.: A novel cryogenic treatment for reduction of residual stresses in 2024 aluminum alloy. Mater. Sci. Eng. A. **689**, 48–52 (2017)
9. McDonach, A., McKelvie, P., MacKenzie, P., Walker, C.A.: Improved Moiré interferometry and applications in fracture mechanics, residual stress and damaged composites. Exp. Tech. **7**(6), 20–24 (1983)
10. Antonov, A.: Development of the method and equipment for holographic inspection of residual-stresses in welded structures. Weld. Prod. **30**, 41–43 (1983)
11. Hung, Y.Y., Ho, H.P.: Shearography: an optical measurement technique and applications. Mater. Sci. Eng. R. Rep. **49**(3), 61–87 (2005)
12. Findeis, D., Gryzagoridis, J.: Determining residual stresses with the aid of optical interference techniques. In: Conference Proceedings of the Society for Experimental Mechanics Series, vol. 4, pp. 277–284 (2013)
13. Peckersky, M.J., Miller, R.F., Vikram, C.S.: Residual stress measurements with laser speckle correlation interferometry and local heat treating. Opt. Eng. **34**(10), 2964–2971 (1995)
14. Viotti, M.R., Albertazzi Jr., A., Kaufmann, G.H.: Measurement of residual stresses using local heating and a radial in-plane speckle interferometer. Opt. Eng. **44**(u), 093606 (2005)
15. Viotti, M.R., Sutério, R., Albertazzi Jr., A., Kaufmann, G.H.: Residual stress measurement using a radial in-plane speckle interferometer and laser annealing: preliminary results. Opt. Lasers Eng. **42**(1), 71–84 (2004)
16. Barile, C., Casavola, C., Pappalettera, G., Pappalettere, C.: Feasibility of local stress relaxation by laser annealing and X-ray measurement. Strain. **49**(5), 393–398 (2013)
17. Barile, C., Casavola, C., Pappalettera, G., Pappalettere, C.: Preliminary analysis for a new approach to relieve residual stresses by laser heating. In: 11th IMEKO TC15 Youth Symposium on Experimental Solid Mechanics 2012, pp. 77–82 (2012)

18. Casavola, C., Pappalettera, G., Pappalettere, C.: ESPI analysis of thermo-mechanical behavior of electronic components. In: Conference Proceedings of the Society for Experimental Mechanics Series, vol. 3,pp. 321–326 (2017)
19. Barile, C., Casavola, C., Pappalettera, G., Pappalettere, C.: Innovative mechanical characterization of materials by combining ESPI and numerical modelling. Int. J. Mech. **10**, 115–123 (2016)
20. Casavola, C., Pappalettera, G., Pappalettere, C.: Design of a double-illumination ESPI system for the measurement of very slow motions. In: Conference Proceedings of the Society for Experimental Mechanics Series, vol. 3, pp. 97–102 (2015)
21. Casavola, C., Lamberti, L., Moramarco, V., Pappalettera, G., Pappalettere, C.: Experimental analysis of thermo-mechanical behaviour of electronic components with speckle interferometry. Strain. **49**(6), 497–506 (2013)
22. Barile, C., Casavola, C., Pappalettera, G., Pappalettere, C.: Hybrid characterization of laminated wood with ESPI and optimization methods. In: Conference Proceedings of the Society for Experimental Mechanics Series, vol. 3, pp. 75–83 (2013)
23. Barile, C., Casavola, C., Pappalettera, G., Pappalettere, C.: Mechanical characterization of SLM specimens with speckle interferometry and numerical optimization. In: Conference Proceedings of the Society for Experimental Mechanics Series, vol. 6, pp. 837–843 (2011)
24. Furgiuele, F.M., Pagnotta, L., Poggialini, A.: Measuring residual stresses by hole-drilling and coherent optics techniques: a numerical calibration. J. Eng. Mater. Technol. **113**(1), 41–50 (1991)
25. Lin, S.T., Hsieh, C.T., Hu, C.P.: Two holographic blind-hole methods for measuring residual stresses. Exp. Mech. **34**(2), 141–147 (1994)
26. Asundi, A., Zhang, J.: Industrial applications of residual stress determination using 2-D in-plane sensitive fibre ESPI and hole-drilling. Proceedings of SPIE, 3740, pp. 78–81 (1999)
27. Lira, I.H., Vial, C., Robinson, K.: The ESPI measurement of the residual stress distribution in chemically etched cold-rolled metallic sheets. Meas. Sci. Technol. **8**(11), 150–157 (1997)
28. Albertazzi, A., Kanda, C., Borges, M.R., Hrebabetzk: A radial in-plane interferometet for ESPI measurement. In: Kujawinska, M., et al. (eds.) Laser Interferometry X: Technique and Analysis Proceedings of SPIE 4101, pp. 77–88 (2002)
29. Suterio, R., Albertazzi, A.G., Cavaco, M.A.M.: Preliminary evaluation: the indentation method combined with a radial interferometer for residual stress measurement. In: Proceedings of the SEM Annual Conference Charlotte (2003)
30. Steinzig, M., Ponslet, E.: Residual stress measurement using the hole drilling method and laser speckle interferometry: part 1. Exp. Tech. **27**(3), 43–46 (2003)
31. Steinzig, M., Ponslet, E.: Residual stress measurement using the hole drilling method and laser speckle interferometry. Exp. Tech. **27**(4), 17–21 (2003)
32. Ponslet, E., Steinzig, M.: Residual stress measurement using the hole drilling method and laser speckle interferometry part III: Analysis. Exp. Tech. **27**(5), 45–48 (2003)
33. Steinzig, M., Takahashi, T.: Residual stress measurement using the hole drilling method and laser speckle interferometry part IV: Measurement accuracy. Exp. Tech. **27**(3), 59–63 (2003)
34. Casavola, C., Pappalettera, G., Pappalettere, C., Tursi, F.: Analysis of the effects of strain measurement errors on residual stresses measured by incremental hole-drilling method. J. Strain Anal. Eng. Des. **48**(5), 313–320 (2013)
35. Rickert, T.J., Gubbels, W.: ESPI hole-drilling of rings and holes using cylindrical hole analysis. In: Conference Proceedings of the Society for Experimental Mechanics Series, vol. 9, pp. 83–89 (2017)
36. Rickert, T.: Stress measurement repeatability in ESPI hole-drilling. In: Conference Proceedings of the Society for Experimental Mechanics Series, vol. 9, pp. 363–369 (2016)
37. Steinzig, M., Upshaw, D., Rasty, J.: Influence of drilling parameters on the accuracy of hole-drilling residual stress measurements. Exp. Mech. **54**(9), 1537–1543 (2014)
38. Barile, C., Casavola, C., Pappalettera, G., Pappalettere, C.: Analysis of the effects of process parameters in residual stress measurements on titanium plates by HDM/ESPI. Measurement. **48**, 220–227 (2014)
39. Barile, C., Casavola, C., Pappalettera, G., Pappalettere, C.: Residual stress measurements by ESPI-HDM in titanium grade 5: comparative measurements with different hole diameters. Ciencia e Tecnologia dos Materiais. **27**(2), 79–93 (2015)
40. Barile, C., Casavola, C., Pappalettera, G., Pappalettere, C.: Considerations on the choice of experimental parameters in residual stress measurements by hole-drilling and ESPI. Frattura ed Integrità Strutturale. **30**, 211–219 (2014)
41. Barile, C., Casavola, C., Pappalettera, G., Pappalettere, C., Tursi, F.: Drilling speed effects on accuracy of HD residual stress measurements. In: Conference Proceedings of the Society for Experimental Mechanics Series, vol. 8, pp. 119–125 (2013)
42. Barile, C., Casavola, C., Pappalettera, G., Pappalettere, C.: Consideration on temperature fields and internal radius of analysis in hdm+espi residual stress measurement. 13th IMEKO TC15 Youth Symposium on Experimental Solid Mechanics, pp. 11–14 (2014)
43. Barile, C., Casavola, C., Pappalettere, C., Pappalettera, G.: Experimental and Numerical Characterization of sintered materials with speckle interferometry and optimization methods. 10th IMEKO TC12 Youth Symposium on Experimental Solid Mechanics, pp. 35–36 (2012)
44. Barile, C., Casavola, C., Pappalettera, G., Pappalettere, C.: Remarks on residual stress measurement by hole-drilling and electronic speckle pattern interferometry. Sci. World J. **2014**, 1–7 (2014)
45. Barile, C., Casavola, C., Pappalettera, G., Pappalettere, C.: Overview of the effects of process parameters on the accuracy in residual stress measurements by using HD and ESPI. In: Conference Proceedings of the Society for Experimental Mechanics Series – Residual Stress, Thermomechanics & Infrared Imaging, Hybrid Techniques and Inverse Problems, vol. 9, pp. 113–118 (2016)
46. Barile, C., Casavola, C., Pappalettera, G., Pappalettere, C.: Residual stress measurement by electronic speckle pattern interferometry; a study of the influence of geometrical parameters. Struct. Integr. Life. **17**(3), 177–182 (2011)
47. Barile, C., Casavola, C., Pappalettera, G., Pappalettere, C.: Residual stress measurement by electronic speckle patterni interferometry; a study of the influence of analysis parameters. Struct. Integr. Life. **12**(3), 159–163 (2012)
48. Baldi, A., Jacquot, P.: Residual stressed investigations in composite samples by speckle interferometry and specimen repositioning. Proceedings SPIE. Speckle Metrol. **4933**, 141–148 (2003)
49. Baldi, A.: A new analytical approach for hole drilling residual stress analysis by full field method. J. Eng. Mater. Technol. **127**, 165–169 (2005)

50. Stefanescu, D., Truman, C.E., Smith, D.J., Whitehead, P.S.: Improvements in residual stress measurement by the incremental centre hole drilling technique. Exp. Mech. **46**(4), 417–427 (2006)
51. Kim, K.S., Choi, S.B., Lee, J.H., Park, S.M., Kim, B.I., Lee, N.H., Lee, C.H., Woo, M.: A study on measurement of welding residual stress using ESPI system. Key Eng. Mater. **324–325**, 859–862 (2006)
52. Lobanov, L.M., Pivtorak, V.A., Savitsky, V.V., Tkachuk, G.I.: Technology and equipment for determination of residual stresses in welded structures based on the application of electron speckle-interferometry. Mater. Sci. Forum. **768–769**, 166–173 (2014)
53. Viotti, M.R., Albertazzi, A.: Compact sensor combining digital speckle pattern interferometry and the hole-drilling technique to measure nonuniform residual stress field. Opt. Eng. **52**(10), 1–8 (2013)
54. Kim, K., Choi, T., Na, M.G., Jung, H.: Residual stress measurement on the butt-welded area by electronic speckle pattern interferometry. Nucl. Eng. Technol. **47**(1), 115–125 (2015)
55. Craft, S., Ghimire, S., Ghimire, B., Yoshida, S., Sasaki, T.: Optical analysis of weld-induced residual stress by electronic speckle-pattern interferometry. In: Conference Proceedings of the Society for Experimental Mechanics Series, vol. 9, pp. 217–323 (2014)
56. Casavola, C., Cazzato, A., Moramarco, V., Pappalettera, G.: Preliminary study on residual stress in FDM parts. In: Conference Proceedings of the Society for Experimental Mechanics Series, vol. 9, pp. 91–96 (2017)
57. Casavola, C., Cazzato, A., Moramarco, V., Pappalettera, G.: Residual stress measurement in fused deposition modelling parts. Polym. Test. **58**, 249–255 (2017)
58. Casavola, C., Cazzato, A., Moramarco, V., Pappalettere, C.: Orthotropic mechanical properties of fused deposition modelling parts described by classical laminate theory. Mater. Des. **90**, 453–458 (2016)
59. Laakkonen, M., Rickert, T., Suominen, L.: Stress measurements in glass and plastic by optical hole-drilling. Mater. Sci. Forum. **768–769**, 95–100 (2014)
60. Takeda, K., Hayashi, S., Ueki, K.: Deformation properties of 3D printed shape memory polymer. Key Eng. Mater. **725**, 378–382 (2017)
61. Montay, G., Sicot, O., Maras, A., Rouhaud, E., Francois, M.: Two dimensions residual stresses analysis through incremental groove machining combined with electronic speckle pattern interferometry. Exp. Mech. **49**, 459–469 (2009)
62. Schajer, G.S., An, Y.: Residual stress determination using cross-slitting and dual-Axis ESPI. Exp. Mech. **50**(2), 169–177 (2010)
63. Schajer, G.S., Steinzig, M.: *Dual-axis hole-drilling ESPI residual stress measurements*. J. Eng. Mater. Technol. Trans. ASME. **132**(1), 0110071–0110075 (2010)
64. Baldi, A.: Combining hole-drilling and ring-core techniques. In: Conference Proceedings of the Society for Experimental Mechanics Series, vol. 9, pp. 105–112 (2017)
65. Schajer, G.S., Rickert, T.J.: Incremental computation technique for residual stress calculations using the integral method. In: Conference Proceedings of the society for Experimental Mechanics Series, vol. 6, pp. 185–191 (2011)
66. Albertazzi, A., Viotti, M.R., Buschinelli, P., Hoffmann, A., Kapp, W.: Residual stress measurement and inner geometry inspection of pipelines by optical methods. In: Conference Proceedings of the Society for Experimental Mechanics Series, vol. 8, pp. 1–12 (2011)
67. Albertazzi, A.G., Viotti, M.R., Kapp, W.A.: A robust achromatic dspi interferometer for measurement in polar coordinates. Tech. Mess. **78**(11), 513–519 (2011)
68. Casavola, C., Pappalettere, C., Tursi, F.: Calibration of barkhausen noise for residual stress measurement. In: Conference Proceedings of the Society for Experimental Mechanics Series, vol. 4, pp. 255–266 (2013)
69. De Paula Dias, A.R., Nunes, R.M., De Lima, T.R.S., Clarke, T.G.R.: Evaluation of the residual stress state of 42crmo4 steel sheets in a production line. Mater. Res. **19**(1), 153–157 (2016)
70. Harrington, J.S., Schajer, G.S.: Measurement of structural stresses by hole-drilling and DIC. Exp. Mech. **57**(4), 1–9 (2017)
71. Baldi, A., Bertolino, F.: A low-cost residual stress measuring instrument. In: Conference Proceedings of the Society for Experimental Mechanics Series, vol 9, pp. 113–119 (2017)
72. Harrington, J., Schajer, G.S.: Measurement of structural stresses by hole-drilling and DIC. In: Conference Proceedings of the Society for Experimental Mechanics, vol. 4, pp. 87–96 (2017)
73. Baldi, A.: Sensitivity analysis of i-DIC approach for residual stress measurement in orthotropic materials. In: Conference Proceedings of the Society for Experimental Mechanics Series, vol. 9, pp. 355–362 (2016)

# Chapter 18
# Feasibility of Using Fringe Projection System for Corrosion Monitoring in Metals of Interest in Cultural Heritage

**C. Casavola, P. Pappalardi, G. Pappalettera, and G. Renna**

**Abstract** Cultural heritage is affected by degradation processes related to intrinsic factors, environmental or human activities. Corrosion is one the most deteriorative phenomena and it can introduce huge damages. Patina or thick corrosion crust formed on artifacts, may display complex products and structures. The identification of the corrosion products or characterization of the patina on artifacts is an essential task to acquire a better knowledge about the condition of ancient objects, corrosion processes and conservation treatment or preventive procedures for long-term, stable preservation. These considerations show that corrosion monitoring is a very important aspect for the evaluation of the degradation of cultural heritage. It is important to have a system that could be used for easy monitoring of the surface condition to put in evidence the beginning of dangerous localized corrosion phenomena. Nowadays optical methods allow detecting and representing, with remarkable accuracy, three-dimensional objects such as sculptures and archaeological finds, as well as large items such as architectural structures. One of the most powerful approaches is the Fringe Projection System (FP) based upon the projection of a complex pattern of light on the object to be reconstructed. In the following work, Fringe Projection is proposed for monitoring and to make quantitative evaluations of the corrosion process occurring on surfaces. The corrosion tests have been performed. Before starting the test, a fringe pattern is projected and the frequency spectrum of the recorded image is analyzed. Successively the samples are submerged in a solution of synthetic acid rain. At different time intervals the specimens were extracted from the solution and analyzed. Fringe pattern is projected again on the surface and new analysis of the frequency spectrum is run. Moreover surface analyses by microscopy and weight loss measurements were performed. At the end of the test, the specimens extracted from the solution were studied by surface analyses and weight loss measurement. The corrosion behavior of the bronze sample has been checked by fringe projection and optical microscopy observation during the exposure to synthetic rain. The average weight loss values were calculated using weight of sample measured before and after exposure to synthetic rain to quantify the corrosion rate of the sample in the corrosive solution.

**Keywords** Bronze Alloy • Corrosion • Cultural Heritage • Fourier Transform • Fringe Projection

## 18.1 Introduction

The cultural heritage is the man's identity. Its preservation and enhancement require an extensive documentation, which includes the physical characteristics of position, shape, color and geometry, and the restoration operations. Nowadays technology based on 3D reconstruction on 3D prototyping can give a great support in restoration planning [1–4] The cultural heritage is affected by degradation processes related to intrinsic factors, environmental or related to human activities, because of their composition and chemical reactivity of the materials that constitute them. These processes alter the appearance and the mechanical and structural characteristics including fatigue behavior [5–12]. Corrosion is one the most detrimental processes [13]. It is a complex chemical-physical phenomenon involving degradation of the materials driven by the environment in which they are exposed. The damage caused by corrosion can be huge. In general the corrosive damage often is not limited to the simple replacement of the affected component (direct costs), but it may include a series of indirect damages. Furthermore, the corrosion is an insidious and a slow process as it displays its negative effects on monthly or yearly base time. One of the most invasive corrosion mechanisms is certainly the one that occurs in marine environments [14, 15]. Another type of very dangerous corrosion is due to exposure to the urban environment, most significantly urban pollution. Several studies have examined the properties of pollutants (water solubility, chemical reactivity, acidity, deposition

C. Casavola • P. Pappalardi • G. Pappalettera (✉) • G. Renna
Dipartimento di Meccanica, Matematica e Management, Politecnico di Bari, Bari, Italy
e-mail: giovanni.pappalettera@poliba.it



L. Lamberti et al. (eds.), *Advancement of Optical Methods in Experimental Mechanics, Volume 3*,
Conference Proceedings of the Society for Experimental Mechanics Series, DOI 10.1007/978-3-319-63028-1_18

velocity) air quality, and the synergistic effects of pollutant mixtures. The conservation state of specific artifacts exposed to polluted urban environments is commonly evaluated by analyzing the patina properties and the degradation of the material [16]. The corrosion of materials occurs with oxidation and the development of an electric current, which goes through the metal to the cathode where the reduction reaction takes place. The result of this process is the consumption of metal and the oxide formation. Patina or thick corrosion crust formed on artifacts, may display complex products and structures. Some of those products or structural details may depend on the microstructure of the metal or the alloy, which is attacked by corrosive agents. Their formation may also be influenced by a variety of growth mechanisms relating to the development and morphology of the corrosion products themselves. The identification of the corrosion products or characterization of the patina on archaeological finds is an essential requisite to acquire a better knowledge about the condition of ancient objects, corrosion processes and conservation treatment or preventive procedures for long-term, stable preservation [17]. Typically, several instruments are used to simulate in the laboratory the action of corrosion on materials such as climate chambers, salt spray and diving alternate (wet and dry), where the materials are subjected to immersion-emersion cycles in a solution [18].

From the above considerations it appears evident that monitoring of the corrosion is a very important aspect for the evaluation of the degradation of cultural heritage artifacts [19]. A relevant aspect that should be considered in developing a proper monitoring system is the possibility to guarantee high sensibility without damaging the measurement object. This indicates that a proper approach should be found among non destructive techniques as well as those based on ultrasounds [20], thermography [21], magnetic noise [22], X-ray [23] etc. In addition, fastness of the analysis and easiness are aspects that should be considered as well [24]. Optical techniques appear to be a promising approach to be explored; in fact they are non-contact techniques that allows to get full-field information with high sensibility that they have been used in many field of experimental mechanics including residual stress measurement [25–30], hybrid testing [31] and slow motion detection [32]. Corrosion onset can be detected as a surface 3D variation so that contouring methods by optical techniques could be implemented to follow the corrosion damage mechanism. One of the most promising approaches for optical contouring is based on Fringe Projection (FP) which essentially consists in analyzing modulation of the fringe pattern projected on the surface to be analyzed [33–37]. The modulated pattern, in fact, contains the information about the object geometry. If compared with laser scanning FP is very effective in reducing the measurement time as it does not require the operation of scanning of the object. By demodulating the information in the recorded pattern it is possible to extract the shape of the examined object, creating a unique match between the image pixels and the points analyzed in the physical space [38]. The determination of the coordinates of the surface of an object is obtained through the Triangulation method [39].

## 18.2 Materials and Methods

In the present study, corrosion was studied on Copper-Tin alloy (Cu-Sn) samples. The chemical composition (wt %) of Cu-Sn alloy is listed in Table 18.1. The material was supplied as plates 1 mm thick: 50 × 100 mm sample were cut from the plate. The specimen surface was mechanically polished by abrasive paper with decreasing granulometry (down to 4000 grit) and degreased with alcohol prior to corrosion test. Corrosion test was performed by immersing the sample for 20 and 110 h in an acid solution (synthetic rain); at the end of each step the sample was dried. The synthetic rain (solution) consists of: 95 ml distilled water, 2.5 ml nitric acid, 1.5 ml hydrochloric acid and 1 ml hydrofluoric acid HF. The surface morphology of the sample after the time exposure to synthetic rain was investigated by Optical Microscope (OM) Nikon Epiphot 200. The weight changes of the sample were measured to a precision of 0.1 mg by analytical balance. The average weight loss results of the sample were calculated using weight of sample measured before and after exposure to synthetic rain. Average weight loss values were applied to calculate the corrosion rate using Eq. (18.1) [40]:

$$\text{Corrosion Rate [mpy]} = \frac{W^{*}3.45^{*}10^{6}}{D^{*}T^{*}A} \tag{18.1}$$

Where the corrosion rate is in mils per year, W is weight loss of the samples given in [g], D is the density of the metal in [g/cm$^3$], T is exposure time in [h] and finally A is the exposure surface area given in [cm$^2$].

**Table 18.1** Chemical composition (wt %) of Cu-Sn alloy

| | Sn | Zn | Pb | Ni | Fe | Sb | P | Cu |
|---|---|---|---|---|---|---|---|---|
| Cu-Sn alloy | 18–21 | 0.5 | 1 | 0.5 | 0.3 | 0.2 | 0.05 | Balance |

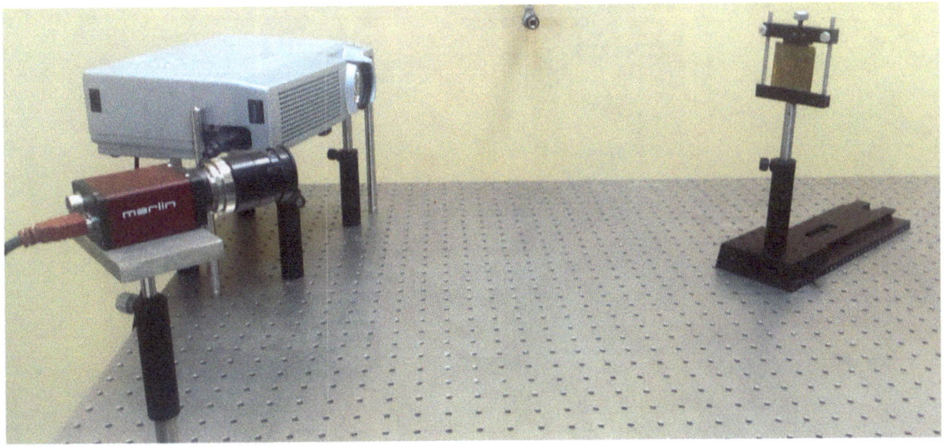


**Fig. 18.1** Picture of the optical fringe projection system

Moreover, at the end of each immersion time, even the specimen images are acquired by a FP system. The optical measuring system is shown in Fig. 18.1. The set-up consists of a standard CCD camera with a resolution of 1392 × 1038 pixels, a frame rate of 17 fps, an ADC of 10 bit and an LCD projector with a resolution of 1024 × 768 pixels. Camera and projector are set in a classical triangulation scheme. The camera is equipped with optics with a focal ratio of 2.8 and a focal length of 50.2 mm. A frame allows to place the object so that it can be fully illuminated by the projector light.

The fixed geometric parameters are the origin of the camera, the distance between the camera and the projector, equal to 530 mm, the angle between the camera and the projector, equal to about 53°, and the distance between the camera and the object. The software tool used to analyze the pattern is implemented in Matlab, while a LabVIEW routine is used to drive the projector. The pattern adopted in this experiment is simply made of vertical stripes with a sinusoidal distribution of gray levels, according Eq. (18.2).

$$I(x,y) = A(x,y) + \sin(2\pi xf + \varphi(x,y)) \tag{18.2}$$

Where, $I$ is the light intensity distribution, $A$ represents the background illumination, $f$ is projection frequency, $\varphi$ is the phase and $x$ in plane horizontal coordinate. The pattern was initially projected on the sample before starting the corrosion process. Successively at different stages during the corrosion test the sample was replaced in the holder and the fringe pattern was projected again on it. For each of the acquired image the 2D FFT was calculated to analyze the frequency content inside the image [41, 42]. Additionally, to check the presence of the corrosion feederate, the surface was examined by optical microscopy and its weight reduction was analyzed.

## 18.3 Results and Discussion

The FFT of the recorded images respectively at the beginning of the test and after 20 and 110 h, are shown in Fig. 18.2. A detail of the frequency content is displayed as recorded after 20 h (a) and 110 h (b) in Fig. 18.3. The reduction of the contrastt can be seen through the decrease in the amplitude spectrum. This is associated with a reduction of the contrast due to the matting of the material, since the advancement of corrosion induces the formation of a patina. Still in Fig. 18.3, in the circles in red, it can be observed the presence of off-axis orders frequencies which were not initially present. This can be associated with the modulation of the projected fringes connected with the formation of the patina, in other words they can be view as indicators of occurring corrosive phenomena.

In Fig. 18.4a the trend of the contrast in correspondence of different frequency orders is plotted. It can be observed that the carrier frequency corresponding to the projected grating decreases as a consequence of the matting of the surface; at the same time the contrast of off-axis orders frequency increase, as shown in Fig. 18.4b.

Also by analyzing (Fig. 18.5) the difference of the FFT spectrum before and after the corrosion process it is possible to observe to presence of new spectral components.

In order to support and confirm the study performed by FP at the end of the tests the surface morphology of sample was analyzed by OM. The surface morphology of bronze sample after 20 and 110 h of exposure to synthetic rain are shown in

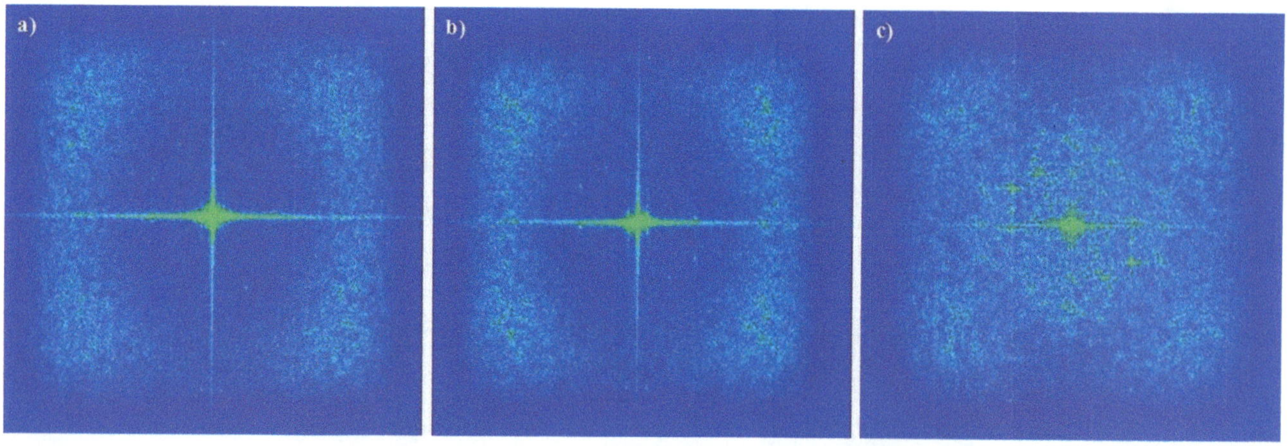


**Fig. 18.2** Images in the frequency domain, respectively to 0 h (**a**) and after 20 h (**b**) and 110 h (**c**) of corrosion test

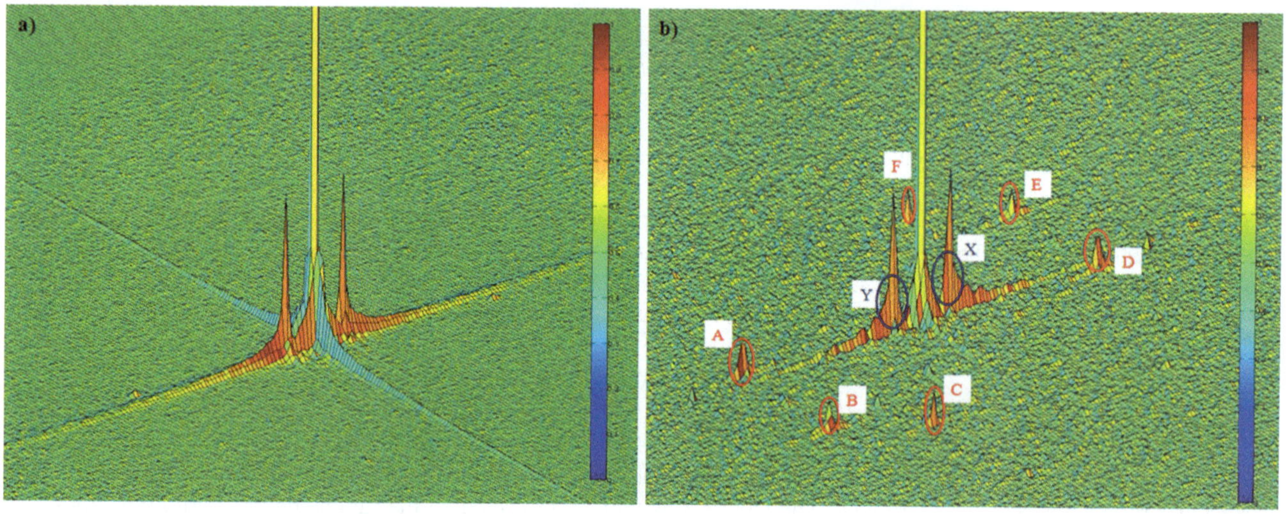


**Fig. 18.3** Detail of the FFT spectrum respectively after 20 h (**a**) and 110 h (**b**) of corrosion test

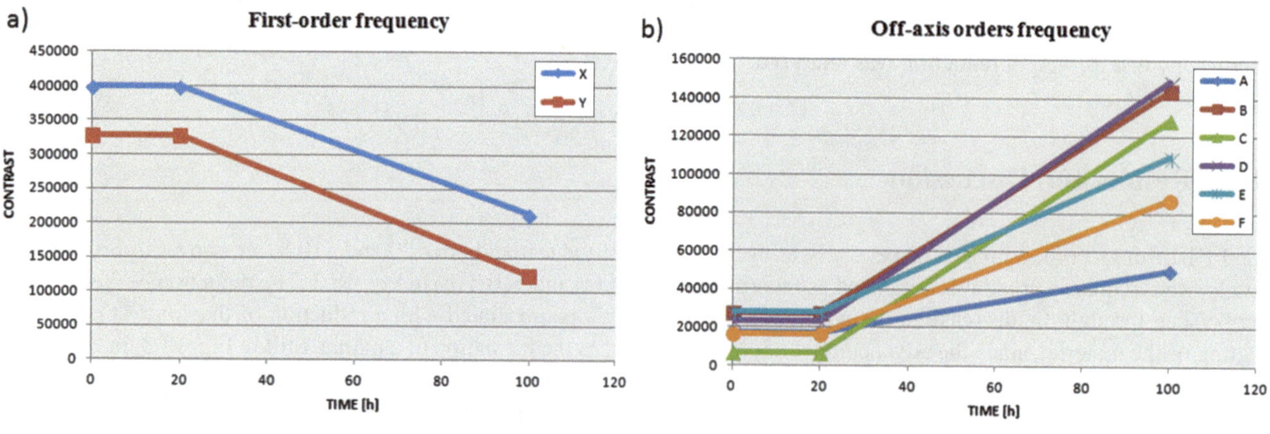


**Fig. 18.4** Trend of the contrast in correspondence of different analyzed frequencies

Fig. 18.6. From a macroscopic point of view, both the surface of sample corroded 20 and 110 h in synthetic rain displays an irregular and inhomogeneous layer of corrosion products. In fact, a different color between the surface exposure to synthetic rain for 20 and 110 h was observed. Also the color on the same surface is not uniform. The color difference variation is due to the patina nature growth and/or dissolution but also to the inhomogeneity of the layers. In particular, the microstructures of the surfaces indicated by the arrows in Fig. 18.6a, b have been shown, at higher magnification, in Fig. 18.7a, b respectively. As

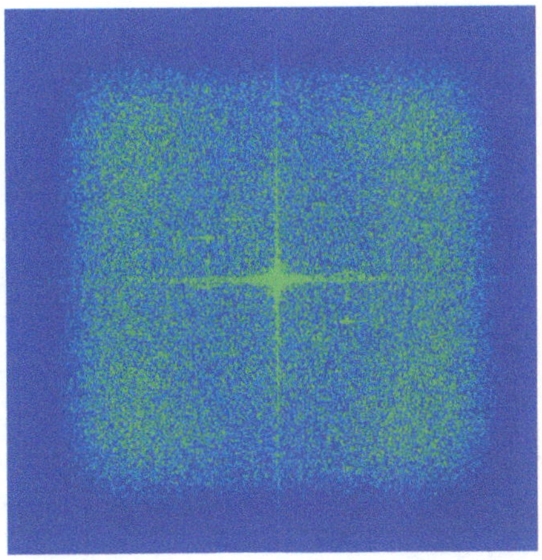

Fig. 18.5 Difference between the Fourier transforms recorded at 0 h and 110 h

Fig. 18.6 Optical macrographs of sample corroded 20 h (**a**), and 110 h in synthetic rain (**b**)

it can be seen the surface of sample is corroded in both cases, especially after 110 h, and the morphologies of the corrosion products are different. On the sample after 20 h some almost perfectly tetragonal crystals can be observed while, on the sample after 110 h it was observed a uniform patina and a preferential corrosion in some areas which correspond to Cu-rich dendrites [43].

In addition, to have information on the general corrosion process the corrosion rate was measured. Corrosion rate calculated using the average weight loss of sample exposed to synthetic rain is shown in Fig. 18.8. The analyses performed on the bronze sample showed that corrosion rate linearly increases with increasing exposure time. This is in full agreement with the sample macrographs observed before.

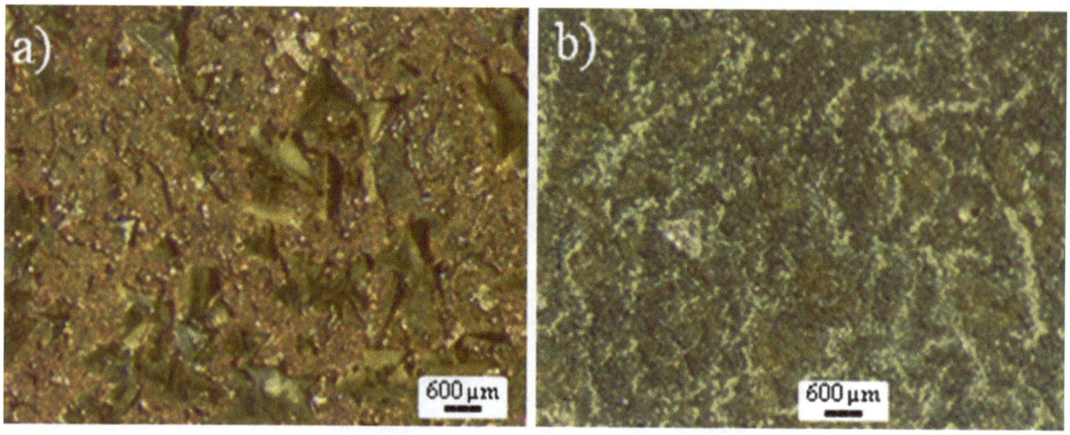


**Fig. 18.7** OM micrographs showing the microstructure of sample corroded 20 h (**a**), and 110 h in synthetic rain (**b**)

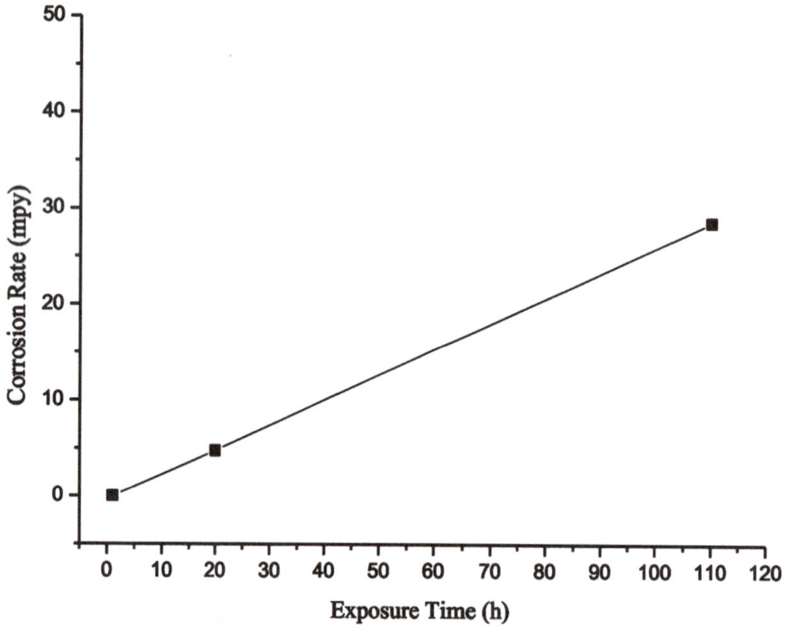


**Fig. 18.8** Corrosion rate as a function of exposure time

## 18.4 Conclusions

Damages introduced by corrosion are huge and, due to this reason, monitoring of the corrosion is a very important aspect for the evaluation of the degradation of materials andthose of interest in cultural heritage. It is important to have a system that can be used for easy monitoring of the surface condition and to put in evidence the beginning of dangerous localized corrosion phenomena, which can require reparation. In this research, a fringe projection system for detecting corrosion onset was tested. By performing a frequency spectrum analysis of the recorded images, in fact, the onset of corrosion determines the contrast reduction of the projected fringe frequency, due of the matting, and the appearance of off-axis orders frequency connected to shape variation introduced by the patina. In particular, it appears that after 20 h of exposure to synthetic rain there are no significant effects while the presence of the patina is evident after 110 h. This preliminary study shows that the approach can have the potential of providing rapid on-site inspection and it could be adapted to measurement of corrosion in a small area or on a large structure.

**Acknowledgements** This research was co-funded by Fondo di Sviluppo e Coesione 2007–2013–APQ Ricerca Regione Puglia "Regional progam for Smart Specialization and Social and Environment sustainability" -FutureInResearch

## References


1. Arbace, L., Sonnino, L., Callieri, M., Dellepiane, M., Fabbri, M., Iaccarino Idelson, A., Scopigno, R.: Innovative uses of 3D digital technologies to assist the resoration of a fragmented terracotta statue. J. Cult. Herit. **14**(4), 332–345 (2013)
2. Fantini, M., de Crescenzio, F., Persiani, F., Benazzi, S., Gruppioni, G.: 3D restitution, restoration and prototyping of a medieval damaged skull. Rapid Prototyp. J. **14**(5), 318–324 (2008)
3. Pieraccini, M., Guidi, G., Atzeni, C.: 3D digitizing of cultural heritage. J. Cult. Herit. **2**(1), 63–70 (2001)
4. Casavola, C., Cazzato, A., Moramarco, V., Pappalettere, C.: Orthotropic mechanical properties of fused deposition modelling parts described by classical laminate theory. Mater. Des. **90**, 453–458 (2016)
5. Chen, Y., Liu, C., Zhou, J., Wang, X.: Multiaxial fatigue behaviors of 2024-T4 aluminium alloy under different corrosion conditions. Int. J. Fatigue. **98**, 269–278 (2017)
6. Casavola, C., Lamberti, L., Pappalettere, C., Tattoli, F.: A comprehensive numerical stress - strain analysis of laser beam butt-welded titanium compared with austenitic steel joints. J. Strain Anal. Eng. Des. **45**(7), 535–554 (2010)
7. Kashani, M.M., Crewe, A.J., Alexander, N.A.: Structural capacity assessment of corroded RC bridge piers. Proceedings of the Institution of Civil Engineers: Bridge Engineering. **170**(1), 28–41 (2017)
8. Casavola, C., Pappalettere, C.: Discussion on local approaches for the fatigue design of welded joints. Int. J. Fatigue. **31**(1), 41–49 (2009)
9. Wei, Z.: Characterization of materials for exhaust systems under combined mechanical and corrosive environment. SAE Technical Paper 9 (2013)
10. Casavola, C., S. Campanelli, L., Pappalettere, C.: Experimental analysis of residual stresses in the selective laser melting process. Society for Experimental Mechanics – 11th International Congress and Exhibition on Experimental and Applied Mechanics, 3, pp. 1479–1486 (2008)
11. Guérin, M., Alexies, J., Andrieu, E., Blanc, C., Odemer, G.: Corrosion-fatigue lifetime of Aluminium-copper-Lithium alloy 2050 in chloride solution. Mater. Des. **87**, 681–692 (2015)
12. Casavola, C., Pappalettere, C.: Application of WEL.FA.RE. method on aluminum alloy welded joints. In: Proceedings of the 2005 SEM Annual Conference and Exposition on Experimental and Applied Mechanics, pp. 1555–1562 (2005)
13. Gaddi, R, Cusano, M, Bonanni, P, Cacace, C, Giovagnoli, A: *Degrado dei Monumenti e Inquinamento Atmosferico: Studio dei Processi di Annerimento, Erosione e Corrosione sui Materiali Costituenti i Beni Culturali di Roma*, Ambiente e Beni Culturali in Collaborazione con ISPRA-ISCR (2015)
14. Poteraş, G., Moncea, M.A., Panait, A.M.: Researches regarding marine environment corrosion on coastal structures. Turk. J. Fish. Aquat. Sci. **14**, 965–971 (2014)
15. Dean, S.W., Delgadillo, G.H.D., Bushman, J.B.: Marine corrosion in tropical environments. Am. Soc. Test. Mater. 75–97 (2000)
16. Gianni, L., Cavallini, M., Natali, S., Adriaens, A.: Wet and dry accelerated aging tests in a spray chamber to understand the effects of acid rain frequencies on bronze corrosion. Int. J. Electrochem. Sci. **8**, 1822–1838 (2013)
17. Ghoniem, M.: The characterization of a corroded Egyptian bronze statue and a study of the degradation phenomena. Int. J. Conserv. Sci. **2**, 95–108 (2011)
18. Chiavari, C., Bernardi, E., Ospitali, F., Robbiola, L., Martini, C., Morselli, L.: *La Corrosione Atmosferica dei Monumenti in Bronzo: Prove di Invecchiamento Artificiale*, La Metallurgia Italiana, pp. 45–54 (2009)
19. Corbellini, S., Ferraris, F., Neri, A., Parvis, M., Angelini, E., Grassini, S.: Exposure-tolerant imaging solution for cultural heritage monitoring. IEEE Trans. Instrum. Meas. **60**(5), 1691–1698 (2011)
20. Moreno Ortiz, F.T., Hernandes Zavaia, A., Gomez-Hernandez, A., Rodriguez Olivares, N.A.: Ultrasonic arc maps and its potential application in non-destructive testing. Conference Proceedings, pp. 111–116 (2017)
21. Cadelano, C., Bortolin, A., Ferrarini, G., Molina, B., Giantin, D., Zonta, P., Bison, P.: Corrosion detection in pipelines using infrared thermography: experiments and data processing methods. J. Nondestruct. Eval. **35**(3), 49 (2016)
22. Casavola, C., Pappalettere, C., Tursi, F.: Calibration of barkhausen noise for residual stress measurement. In: Conference Proceedings of the Society for Experimental Mechanics Series, vol. 4, pp. 255–266 (2013)
23. Barile, C., Casavola, C., Pappalettera, G., Pappalettere, C.: Feasibility of local stress relaxation by laser annealing and x-ray measurement. Strain. **49**(5), 393–398 (2013)
24. Huang, P.S., Jin, F., Chiang, F.P.: Quantitative evaluation of corrosion by a digital fringe projection technique. Opt. Lasers Eng. **31**, 371–380 (1999)
25. Barile, C., Casavola, C., Pappalettera, G., Pappalettere, C., Tursi, F.: Drilling speed effects on accuracy of HD residual stress measurements. In: Conference Proceedings of the Society for Experimental Mechanics Series, vol. 8, pp. 119–125 (2014)
26. McDonach, A., McKelvie, P., MacKenzie, P., Walker, C.A.: Improved Moiré interferometry and applications in fracture mechanics, residual stress and damaged composites. Exp. Tech. **7**(6), 20–24 (1983)
27. Barile, C., Casavola, C., Pappalettera, G., Pappalettere, C.: Mechanical characterization of SLM specimens with speckle interferometry and numerical optimization. In: Conference Proceedings of the Society for Experimental Mechanics Series, vol. 6, pp. 837–843 (2011)
28. Hung, Y.Y., Ho, H.P.: Shearography: an optical measurement technique and applications. Mater. Sci. Eng. R. Rep. **49**(3), 61–87 (2005)
29. Schajer, G.S., Steinzig, M.: Dual-axis hole-drilling ESPI residual stress measurements. J. Eng. Mater. Technol. Trans. ASME. **132**(1), 0110071–0110075 (2010)
30. Baldi, A.: Combining hole-drilling and ring-core techniques. In: Conference Proceedings of the Society for Experimental Mechanics Series, vol. 9, pp. 105–112 (2017)
31. Barile, C., Casavola, C., Pappalettera, G., Pappalettere, C.: Hybrid characterization of laminated wood with ESPI and optimization methods. In: Conference Proceedings of the Society for Experimental Mechanics Series, vol. 3, pp. 75–83 (2013)
32. Casavola, C., Pappalettera, G., Pappalettere, C.: Design of a Double-Illumination ESPI system for the measurement of very slow motions. In: Conference Proceedings of the Society for Experimental Mechanics Series, vol. 3, pp. 97–102 (2014)
33. Miao, H., Quan, C., Tay, C.J., Fu, Y., Wu, X.P.: Optical edge projection for surface contouring. Opt. Commun. **256**, 16–23 (2005)

34. Casavola, C., Lamberti, L., Pappalettera, G., Pappalettere, C.: Application of contouring to dental reconstruction. In: Conference Proceedings of the Society for Experimental Mechanics Series, vol. 3, pp. 183–191 (2013)
35. Shi, Z., Lin, J.: Method for 3D profilometry measurement based on contouring moire fringes. Proceedings of SPIE 7723, article number 672302 (2007)
36. Casavola, C., Pappalettera, G., Pappalettere, C.: Design of a fiber optics fringe projector for 3D reconstruction of dental elements. 3rd Mediterranean Photonics Conference, article number 5866487 (2014)
37. Miao, H., Quan, C., Tay, C.J., Wu, X.P.: Profilometry using optical edge projection. Proc. SPIE. **5852**, 169–174 (2004)
38. Broggiato, G.B., Campana, F., Gerbino, S., Martorelli, M.: *Confronto tra Diverse Tecniche di Digitalizzazione delle Forme per il Reverse Engineering*, INGEGRAF, pp. 1–11 (2002)
39. Sitnik, R.: A fully automatic 3D shape measurement system with data export for engineering and multimedia systems. A Dissertation Submitted in Partial Fulfillment of the Requirements for the Degree of Doctor of Philosophy in the Warsaw University of Technology Warsaw (2002)
40. Norouzi, S., Eslami, F., Wyszynski, M.L., Tsolakis, A.: Corrosion effects of RME in blends with ULSD on aluminum and copper. Fuel Process. Technol. **104**, 204–210 (2012)
41. Bogrekci, I., Durakbasa, N., Demircioglu, P.: Comparison between 3D digital and optical microscopes for the surface measurement using image processing techniques. Automatisierungstechnik. **61**(3), 195–202 (2013)
42. Spagnolo, G.S., Guattari, G., Sapia, C., Ambrosini, D., Paoletti, D., Accardo, G.: Contouring of artwork surface by fringe projection and FFT analysis. Opt. Lasers Eng. **33**, 141–156 (2000)
43. Bernardi, E., Chiavari, C., Martini, C., Morselli, L.: The atmospheric corrosion of quaternary bronzes: an evaluation of the dissolution rate of the alloying elements. Appl. Phys. A. **92**(1), 83–89 (2008)

FSC
www.fsc.org
MIX
Papier | Fördert gute Waldnutzung
FSC® C083411

Kolibri 360 GmbH
Ferdinand-Jühlke-Straße 7
99095 Erfurt, Deutschland
produktsicherheit@kolibri360.de